THEORY OF HEAT.

BY

J. CLERK MAXWELL, M.A.

LL.D. EDIN., F.R.SS. L. & E.

Honorary Fellow of Trinity College and Professor of Experimental Physics in the University of Cambridge.

THIRD EDITION.

GREENWOOD PRESS, PUBLISHERS
WESTPORT, CONNECTICUT

Originally published in 1872
by Longmans, Green and Company, London

First Greenwood Reprinting 1970

Library of Congress Catalogue Card Number 69-13993

SBN 8371-4097-8

Printed in the United States of America

PREFACE.

THE AIM of this book is to exhibit the scientific connexion of the various steps by which our knowledge of the phenomena of heat has been extended. The first of these steps is the invention of the thermometer, by which the registration and comparison of temperatures is rendered possible. The second step is the measurement of quantities of heat, or Calorimetry. The whole science of heat is founded on Thermometry and Calorimetry, and when these operations are understood we may proceed to the third step, which is the investigation of those relations between the thermal and the mechanical properties of substances which form the subject of Thermodynamics. The whole of this part of the subject depends on the consideration of the Intrinsic Energy of a system of bodies, as depending on the temperature and physical state, as well as the form, motion, and relative position of these bodies. Of this energy, however, only a part is available for the purpose of producing mechanical work, and though the energy itself is indestructible, the available part is liable to diminution by the action of certain natural processes, such as conduction and radiation of heat, friction, and viscosity. These processes, by which energy is rendered unavailable as a source of work, are classed together under the name of the Dissipation of Energy, and form the subjects of the next division of the book. The last chapter is devoted to the explanation of various phenomena by means of the hypothesis that bodies consist of molecules, the motion of which constitutes the heat of those bodies.

In order to bring the treatment of these subjects within the limits of this text-book, it has been found necessary to omit everything which is not an essential part of the intellectual process by which the doctrines of heat have been developed, or which does not materially assist the student in forming his own judgment on these doctrines.

For this reason, no account is given of several very important experiments, and many illustrations of the theory of heat by means of natural phenomena are omitted. The student, however, will find this part of the subject treated at greater length in several excellent works on the same subject which have lately appeared.

A full account of the most important experiments on the effects of heat will be found in Dixon's 'Treatise on Heat' (Hodges & Smith, 1849).

Professor Balfour Stewart's treatise contains all that is necessary to be known in order to make experiments on heat. The student may be also referred to Deschanel's 'Natural Philosophy,' Part II., translated by Professor Everett, who has added a chapter on Thermodynamics; to Professor Rankine's work on the Steam Engine, in which he will find the first systematic treatise on thermodynamics; to Professor Tait's 'Thermodynamics,' which contains an historical sketch of the subject, as well as the mathematical investigations; and to Professor Tyndall's work on 'Heat as a Mode of Motion,' in which the doctrines of the science are forcibly impressed on the mind by well-chosen illustrative experiments. The original memoirs of Professor Clausius, one of the founders of the modern science of Thermodynamics, have been edited in English by Professor Hirst.

CONTENTS.

CHAPTER I

INTRODUCTION.

CHAPTER II.

THERMOMETRY, OR THE REGISTRATION OF TEMPERATURE.

CHAPTER III.

CALORIMETRY, OR THE MEASUREMENT OF HEAT.

CHAPTER IV.

ELEMENTARY DYNAMICAL PRINCIPLES.

CHAPTER V.

MEASUREMENT OF INTERNAL FORCES AND THEIR EFFECTS.

CHAPTER VI.

LINES OF EQUAL TEMPERATURE ON THE INDICATOR DIAGRAM.

CHAPTER VII.

ADIABATIC LINES.

CHAPTER VIII.

HEAT ENGINES.

CHAPTER IX.

RELATIONS BETWEEN THE PHYSICAL PROPERTIES OF A SUBSTANCE.

CHAPTER X.

LATENT HEAT.

CHAPTER XI.

THERMODYNAMICS OF GASES.

CHAPTER XII.

ON THE INTRINSIC ENERGY OF A SYSTEM OF BODIES.

CHAPTER XIII.

ON FREE EXPANSION.

CHAPTER XIV.

DETERMINATION OF HEIGHTS BY THE BAROMETER.

CHAPTER XV.

ON THE PROPAGATION OF WAVES OF LONGITUDINAL DISTURBANCE.

CHAPTER XVI.

ON RADIATION.

CHAPTER XVII.

ON CONVECTION CURRENTS.

CHAPTER XVIII.

ON THE DIFFUSION OF HEAT BY CONDUCTION.

CHAPTER XIX.

ON THE DIFFUSION OF FLUIDS.

CHAPTER XX.

ON CAPILLARITY.

CHAPTER XXI.

ON ELASTICITY AND VISCOSITY.

CHAPTER XXII.

MOLECULAR THEORY OF THE CONSTITUTION OF BODIES.

A TREATISE

ON

HEAT.

CHAPTER I.

INTRODUCTION.

THE DISTINCTION between hot bodies and cold ones is familiar to all, and is associated in our minds with the difference of the sensations which we experience in touching various substances, according as they are hot or cold. The intensity of these sensations is susceptible of degrees, so that we may estimate one body to be hotter or colder than another by the touch. The words hot, warm, cool, cold, are associated in our minds with a series of sensations which we suppose to indicate a corresponding series of states of an object with respect to heat.

We use these words, therefore, as the names of these states of the object, or, in scientific language, they are the names of Temperatures, the word hot indicating a high temperature, cold a low temperature, and the intermediate terms intermediate temperatures, while the word temperature itself is a general term intended to apply to any one of these states of the object.

Since the state of a body may vary continuously from cold to hot, we must admit the existence of an indefinite number of intermediate states, which we call intermediate temperatures. We may give names to any number of particular degrees of temperature, and express any other temperature by its relative place among these degrees.

The temperature of a body, therefore, is a quantity which indicates how hot or how cold the body is.

When we say that the temperature of one body is higher or lower than that of another, we mean that the first body is hotter or colder than the second, but we also imply that we refer the state of both bodies to a certain scale of temperatures. By the use, therefore, of the word temperature, we fix in our minds the conviction that it is possible, not only to feel, but to measure, how hot a body is.

Words of this kind, which express the same things as the words of primitive language, but express them in a way susceptible of accurate numerical statement, are called scientific[1] terms, because they contribute to the growth of science.

We might suppose that a person who has carefully cultivated his senses would be able by simply touching an object to assign its place in a scale of temperatures, but it is found by experiment that the estimate formed of temperature by the touch depends upon a great variety of circumstances, some of these relating to the texture or consistency of the object, and some to the temperature of the hand or the state of health of the person who makes the estimate.

For instance, if the temperature of a piece of wood were the same as that of a piece of iron, and much higher than that of the hand, we should estimate the iron to be hotter than the wood, because it parts with its heat more readily to the hand, whereas if their temperatures were equal, and much lower than that of the hand, we should estimate the iron to be colder than the wood.

There is another common experiment, in which we place one hand in hot water and the other in cold for a sufficient

[1] 'Scientifick, adj. Producing demonstrative knowledge.'—*Johnson's Dict.*

time. If we then dip both hands in the same basin of lukewarm water alternately, or even at once, it will appear cold to the warmed hand and hot to the cooled hand.

In fact, our sensations of every kind depend upon so many variable conditions, that for all scientific purposes we prefer to form our estimate of the state of bodies from their observed action on some apparatus whose conditions are more simple and less variable than those of our own senses.

The properties of most substances vary when their temperature varies. Some of these variations are abrupt, and serve to indicate particular temperatures as points of reference; others are continuous, and serve to measure other temperatures by comparison with the temperatures of reference.

For instance, the temperature at which ice melts is found to be always the same under ordinary circumstances, though, as we shall see, it is slightly altered by change of pressure. The temperature of steam which issues from boiling water is also constant when the pressure is constant.

These two phenomena therefore—the melting of ice and the boiling of water—indicate in a visible manner two temperatures which we may use as points of reference, the position of which depends on the properties of water and not on the conditions of our senses.

Other changes of state which take place at temperatures more or less definite, such as the melting of wax or of lead, and the boiling of liquids of definite composition, are occasionally used to indicate when these temperatures are attained, but the melting of ice and the boiling of pure water at a standard pressure remain the most important temperatures of reference in modern science.

These phenomena of change of state serve to indicate only a certain number of particular temperatures. In order to measure temperatures in general, we must avail ourselves of some property of a substance which alters continuously with the temperature.

The volume of most substances increases continuously as the temperature rises, the pressure remaining constant. There are exceptions to this rule, and the dilatations of different substances are not in general in the same proportion; but any substance in which an increase of temperature, however small, produces an increase of volume may be used to indicate changes of temperature.

For instance, mercury and glass both expand when heated, but the dilatation of mercury is greater than that of glass. Hence if a cold glass vessel be filled with cold mercury, and if the vessel and the mercury in it are then equally heated, the glass vessel will expand, but the mercury will expand more, so that the vessel will no longer contain the mercury. If the vessel be provided with a long neck, the mercury forced out of the vessel will rise in the neck, and if the neck is a narrow tube finely graduated, the amount of mercury forced out of the vessel may be accurately measured.

This is the principle of the common mercurial thermometer, the construction of which will be afterwards more minutely described. At present we consider it simply as an instrument the indications of which vary when the temperature varies, but are always the same when the temperature of the instrument is the same.

The dilatation of other liquids, as well as that of solids and of gases, may be used for thermometric purposes, and the thermo-electric properties of metals, and the variation of their electric resistance with temperature, are also employed in researches on heat. We must first, however, study the theory of temperature in itself before we examine the properties of different substances as related to temperature, and for this purpose we shall use the particular mercurial thermometer just described.

THE MERCURIAL THERMOMETER.

This thermometer consists of a glass tube terminating in a bulb, the bulb and part of the tube being filled with mercury, and the rest of the tube being empty. We shall suppose the tube to be graduated in any manner so that the height of the mercury in the tube may be observed and recorded. We shall not, however, assume either that the tube is of uniform section or that the degrees are of equal size, so that the scale of this primitive thermometer must be regarded as completely arbitrary. By means of our thermometer we can ascertain whether one temperature is higher or lower than another, or equal to it, but we cannot assert that the difference between two temperatures, A and B, is greater or less than the difference between two other temperatures, C and D.

We shall suppose that in every observation the temperature of the mercury and the glass is equal and uniform over the whole thermometer. The reading of the scale will then depend on the temperature of the thermometer, and, since we have not yet established any more perfect thermometric scale, we shall call this reading provisionally 'the temperature by the arbitrary scale of the thermometer.'

The reading of a thermometer indicates primarily its own temperature, but if we bring the thermometer into intimate contact with another substance, as for instance if we plunge it into a liquid for a sufficient time, we find that the reading of the thermometer becomes higher or lower according as the liquid is hotter or colder than the thermometer, and that if we leave the thermometer in contact with the substance for a sufficient time the reading becomes stationary. Let us call this ultimate reading 'the temperature of the substance.' We shall find as we go on that we have a right to do so.

Let us now take a vessel of water which we shall suppose to be at the temperature of the air, so that if left to itself it would remain at the same temperature. Take another smaller vessel of thin sheet copper or tin plate, and fill it with water, oil, or any other liquid, and immerse it in the larger vessel of water for a certain time. Then, if by means of our thermometer we register the temperatures of the liquids in the two vessels before and after the immersion of the copper vessel, we find that if they are originally at the same temperature the temperature of both remains the same, but that if one is at a higher temperature than the other, that which has the higher temperature becomes colder and that which has the lower temperature becomes hotter, so that if they continue in contact for a sufficient time they arrive at last at the same temperature, after which no change of temperature takes place.

The loss of temperature by the hot body is not in general equal to the gain of temperature by the cold body, but it is manifest that the two simultaneous phenomena are due to one cause, and this cause may be described as the passage of Heat from the hot body to the cold one.

As this is the first time we have used the word Heat, let us examine what we mean by it.

We find the cooling of a hot body and the heating of a cold body happening simultaneously as parts of the same phenomenon, and we describe this phenomenon as the passage of heat from the hot body to the cold one. Heat, then, is something which may be transferred from one body to another, so as to diminish the quantity of heat in the first and increase that in the second by the same amount. When heat is communicated to a body, the temperature of the body is generally increased, but sometimes other effects are produced, such as change of state. When heat leaves a body, there is either a fall of temperature or a change of state. If no heat enters or leaves a body, and if no changes of state or mechanical actions take place in the body, the temperature of the body will remain constant.

Heat, therefore, may pass out of one body into another just as water may be poured from one vessel into another, and it may be retained in a body for any time, just as water may be kept in a vessel. We have therefore a right to speak of heat as of a *measurable quantity*, and to treat it mathematically like other measurable quantities so long as it continues to exist as heat. We shall find, however, that we have no right to treat heat as a *substance*, for it may be transformed into something which is not heat, and is certainly not a substance at all, namely, mechanical work.

We must remember, therefore, that though we admit heat to the title of a measurable quantity, we must not give it rank as a substance, but must hold our minds in suspense till we have further evidence as to the nature of heat.

Such evidence is furnished by experiments on friction, in which mechanical work, instead of being transmitted from one part of a machine to another, is apparently lost, while at the same time, and in the same place, heat is generated, the amount of heat being in an exact proportion to the amount of work lost. We have, therefore, reason to believe that heat is of the same nature as mechanical work, that is, it is one of the forms of Energy.

In the eighteenth century, when many new facts were being discovered relating to the action of heat on bodies, and when at the same time great progress was being made in the knowledge of the chemical action of substances, the word Caloric was introduced to signify heat as a measurable quantity. So long as the word denoted nothing more than this, it might be usefully employed, but the form of the word accommodated itself to the tendency of the chemists of that time to seek for new 'imponderable substances,' so that the word caloric came to *connote*[1] not merely heat, but heat as an indestructible imponderable fluid, insinuating itself into the pores of bodies, dilating and dissolving them, and

[1] 'A connotative term is one which denotes a subject and implies an attribute.'—*Mill's Logic*, book i. chap. ii. § 5.

ultimately vaporising them, combining with bodies in definite quantities, and so becoming latent, and reappearing when these bodies alter their condition. In fact, the word caloric, when once introduced, soon came to imply the recognised existence of something material, though probably of a more subtle nature than the then newly discovered gases. Caloric resembled these gases in being invisible and in its property of becoming fixed in solid bodies. It differed from them because its weight could not be detected by the finest balances, but there was no doubt in the minds of many eminent men that caloric was a fluid pervading all bodies, probably the cause of all repulsion, and possibly even of the extension of bodies in space.

Since ideas of this kind have always been connected with the word caloric, and the word itself has been in no slight degree the means of embodying and propagating these ideas, and since all these ideas are now known to be false, we shall avoid as much as possible the use of the word caloric in treating of heat. We shall find it useful, however, when we wish to refer to the erroneous theory which supposes heat to be a substance, to call it the 'Caloric Theory of Heat.'

The word heat, though a primitive word and not a scientific term, will be found sufficiently free from ambiguity when we use it to express a measurable quantity, because it will be associated with words expressive of quantity, indicating how much heat we are speaking of.

We have nothing to do with the word heat as an abstract term signifying the property of hot things, and when we might say a certain heat, as the heat of new milk, we shall always use the more scientific word temperature, and speak of the temperature of new milk.

We shall never use the word heat to denote the sensation of heat. In fact, it is never so used in ordinary language, which has no names for sensations, unless when the sensation itself is of more importance to us than its physical cause, as in the case of pain, &c. The only name we have for this sensation is 'the sensation of heat.'

When we require an adjective to denote that a phenomenon is related to heat we shall call it a *thermal* phenomenon, as, for instance, we shall speak of the thermal conductivity of a substance or of thermal radiation to distinguish the conduction and radiation of heat from the conduction of electricity or the radiation of light. The science of heat has been called (by Dr. Whewell and others) Thermotics, and the theory of heat as a form of energy is called Thermodynamics. In the same way the theory of the equilibrium of heat might be called Thermostatics, and that of the motion of heat Thermokinematics.

The instrument by which the temperature of bodies is registered is called a Thermometer or measurer of warmth, and the method of constructing and using thermometers may be called Thermometry.

The instrument by which quantities of heat are measured is called a Calorimeter, probably because it was invented at a time when heat was called Caloric. The name, however, is now well established, and is a convenient one, as its form is sufficiently distinct from that of the word Thermometer. The method of measuring heat may be called Calorimetry.

A certain quantity of heat, with which all other quantities are compared, is called a Thermal Unit. This is the quantity of heat required to produce a particular effect, such as to melt a pound of ice, or to raise a pound of water from one defined temperature to another defined temperature. A particular thermal unit has been called by some authors a Calorie.

We have now obtained two of the fundamental ideas of the science of heat—the idea of temperature, or the property of a body considered with reference to its power of heating other bodies; and the idea of heat as a measurable quantity, which may be transferred from hotter bodies to colder ones. We shall consider the further development of these ideas in the chapters on Thermometry and Calorimetry, but we must first direct our attention to the process by which heat is transferred from one body to another.

This process is called the Diffusion of Heat. The diffusion of heat invariably transfers heat from a hotter body to a colder one, so as to cool the hotter body and warm the colder body. This process would go on till all bodies were brought to the same temperature if it were not for certain other processes by which the temperatures of bodies are changed independently of any exchange of heat with other bodies, as, for instance, when combustion or any other chemical process takes place, or when any change occurs in the form, structure, or physical state of the body.

The changes of temperature of a body arising from other causes than the transfer of heat from other bodies will be considered when we come to describe the different physical states of bodies. We are at present concerned only with the passage of heat into the body or out of it, and this always takes place by diffusion, and is always from a hotter to a colder body.

Three processes of diffusion of heat are commonly recognised—Conduction, Convection, and Radiation.

Conduction is the flow of heat through an unequally heated body from places of higher to places of lower temperature.

Convection is the motion of the hot body itself carrying its heat with it. If by this motion it is brought near bodies colder than itself it will warm them faster than if it had not been moved nearer to them. The term convection is applied to those processes by which the diffusion of heat is rendered more rapid by the motion of the hot substance from one place to another, though the ultimate transfer of heat may still take place by conduction.

In Radiation, the hotter body loses heat, and the colder body receives heat by means of a process occurring in some intervening medium which does not itself become thereby hot.

In each of these three processes of diffusion of heat the temperatures of the bodies between which the process takes place tend to become equal. We shall not at present discuss the convection of heat, because it is not a purely thermal phenomenon, since it depends on a hot substance being carried from one place to another, either by human effort, as when a hot iron is taken out of the fire and put into the tea-urn, or by some natural property of the heated substance, as when water, heated by contact with the bottom of a kettle placed on the fire, expands as it becomes warmed, and forms an ascending current, making way for colder and therefore denser water to descend and take its place. In every such case of convection the ultimate and only direct transfer of heat is due to conduction, and the only effect of the motion of the hot substance is to bring the unequally heated portions nearer to each other, so as to facilitate the exchange of heat. We shall accept the conduction of heat as a fact, without at present attempting to form any theory of the details of the process by which it takes place. We do not even assert that in the diffusion of heat by conduction the transfer of heat is entirely from the hotter to the colder body. All that we assert is, that the amount of heat transferred from the hotter to the colder body is invariably greater than the amount, if any, transferred from the colder to the hotter.

ON CONDUCTION.

In the experiments which we have described, heat passes from one body into another through an intervening substance, as from a vessel of water through the glass bulb of a thermometer into the mercury inside the bulb.

This process, by which heat passes from hotter to colder parts of a body, is called the conduction of heat. When heat is passing through a body by conduction, the temperature of the body must be greater in the parts from which the heat comes than in those to which it tends, and the quantity of heat which passes through any thin layer of the substance depends on the difference of the temperatures of the opposite sides of the layer. For instance, if we put a silver spoon into a cup of hot tea, the part of the spoon in the tea soon becomes heated, while the part just out of the tea is comparatively cool. On account of this inequality of temperature, heat immediately begins to flow along the metal from A to B. The heat first warms B a little, and so makes B warmer than C, and then the heat flows on from B to C, and in this way the very end of the spoon will in course of time become warm to the touch. The essential requisite to the conduction of heat is, that in every part of its course the heat must pass from hotter to colder parts of the body. No heat can be conducted as far as E till A has been made hotter than B, B than C, C than D, and D than E. To do this requires a certain amount of heat to be expended in warming in succession all these intermediate parts of the spoon, so that for some time after the spoon is placed in the cup no alteration of temperature can be perceived at the end of the spoon.

Fig. 1.

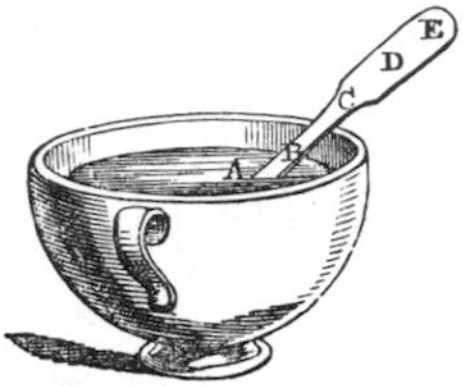

Hence we may define conduction as the passage of heat through a body depending on inequality of temperature in adjacent parts of the body.

When any part of a body is heated by conduction, the parts of the body through which the heat comes to it must be hotter than itself, and the parts higher up the stream of heat still hotter.

If we now try the experiment of the spoon in the teacup with a German silver spoon along with the silver one, we shall find that the end of the silver spoon becomes hot long before that of the German silver one; and if we also put in a bone or horn spoon, we shall not be able to perceive any warmth at the end of it, however long we wait.

This shows that silver conducts heat quicker than German

silver, and German silver quicker than bone or horn. The reason why the end of the spoon never gets as hot as the tea is, that the intermediate parts of the spoon are cooling, partly by giving their heat to the air in contact with them, and partly by radiation out into space.

To show that the first effect of heat on the thermometer is to warm the material of which the bulb is composed, and that the heat cannot reach the fluid inside till the bulb has been warmed, take a thermometer with a large bulb, watch the fluid in the tube, and dash a little hot water over the bulb. The fluid will fall in the tube before it begins to rise, showing that the bulb began to expand before the fluid expanded.

ON RADIATION.

On a calm day in winter we feel the sun's rays warm even when water is freezing and ice is hard and dry.

If we make use of a thermometer, we find that if the sun's rays fall on it, it indicates a temperature far above freezing, while the air immediately surrounding the bulb is at a temperature below freezing. The heat, therefore, which we feel, and to which the thermometer also responds, is not conveyed to it by conduction through the air, for the air is cold, and a cold body cannot make a body warmer than itself by conduction. The mode in which the heat reaches the body which it warms, without warming the air through which it passes, is called radiation. Substances which admit of radiation taking place through them are called Diathermanous. Those which do not allow heat to pass through them without becoming themselves hot are called Athermanous. That which passes through the medium during this process is generally called Radiant Heat, though as long as it is radiant it possesses none of the properties which distinguish heat from other forms of energy, since the temperature of the body through which it passes, and the other physical properties of the body, are in no way affected by the passage of the radiation, provided the body is perfectly diathermanous. If the body is not perfectly diathermanous it stops more or less of the radiation, and becomes heated itself, instead of transmitting the whole radiation to bodies beyond it.

The distinguishing characteristic of radiant heat is, that it travels in *rays* like light, whence the name radiant. These rays have all the physical properties of rays of light, and are capable of reflexion, refraction, interference, and polarisation. They may be divided into different kinds by the prism, as light is divided into its component colours, and some of the heat-rays are identical with the rays of light, while other kinds of heat-rays make no impression on our eyes. For instance, if we take a glass convex lens, and place it in the sun's rays, a body placed at the focus where a small image of the sun is formed will be intensely heated, while the lens itself and the air through which the rays pass remain quite cold. If we allow the rays before they reach the focus to fall on the surface of water, so that the rays meet in a focus in the interior of the water, then if the water is quite clear at the focus it will remain tranquil, but if we make the focus fall upon a mote in the water, the rays will be stopped, the mote will be heated and will cause the water next it to expand, and so an upward current will be produced, and the mote will begin to rise in the water. This shows that it is only when the radiation is *stopped* that it has any effect in heating what it falls on.

By means of any regular concave piece of metal, such as the scale of a balance, pressed when hot against a clear sheet of ice, first on one side and then on the other, it is easy to make a lens of ice which may be used on a sunny day as a burning glass; but this experiment, which was formerly in great repute, is far inferior in interest to one invented by Professor Tyndall, in which the heat, instead of being concentrated *by* ice, is concentrated *in* ice. Take a clear block of ice and make a flat surface on it, parallel to the original surface of the lake, or to the layers of bubbles generally found in large blocks; then let the converging rays of the sun from an ordinary burning glass fall on this surface, and come to a focus within the ice. The ice, not being perfectly diathermanous, will be warmed by the rays, but much more at the focus than anywhere else. Thus the ice will begin to melt at the focus in the interior of its substance, and, as it does so, the portions which melt first are regularly formed crystals, and so we see in the path of the beam a number of six-rayed stars, which are hollows cut out of the ice and containing water. This water, however, does not quite fill them, because the water is of less bulk than the ice of which it was made, so that parts of the stars are empty.

Experiments on the heating effects of radiation show that not only the sun but all hot bodies emit radiation. When the body is hot enough, its radiations become visible, and the body is said to be red hot. When it is still hotter it sends forth not only red rays, but rays of every colour, and it is then said to be white hot. When a body is too cold to shine visibly, it still shines with invisible heating rays, which can be perceived by a sufficiently delicate thermometer, and it does not appear that any body can be so cold as not to send forth radiations. The reason why all bodies do not appear to shine is, that our eyes are sensitive only to particular kinds of rays, and we only see by means of rays of these kinds, coming from some very hot body, either directly or after reflexion or scattering at the surface of other bodies.

We shall see that the phrases radiation of heat and radiant heat are not quite scientifically correct, and must be used with caution. Heat is certainly communicated from one body to another by a process which we call radiation, which takes place in the region between the two bodies. We have no right, however, to speak of this process of radiation as heat. We have defined heat as it exists in hot bodies, and we have seen that all heat is of the same kind. But the radiation between bodies differs from heat as we have defined it—1st, in not making the body hot through which it passes; 2nd, in being of many different kinds. Hence we shall generally speak of radiation, and when we speak of radiant heat we do not mean to imply the existence of a new kind of heat, but to consider radiation in its thermal aspect.

ON THE DIFFERENT PHYSICAL STATES OF BODIES.

Bodies are found to behave in different ways under the action of forces. If we cause a longitudinal pressure to act on a body in one direction by means of a pair of pincers or a vice, the body being free to move in all other directions, we find that if the body is a piece of cold iron there is very little effect produced, unless the pressure be very great; if the body is a piece of india-rubber, it is compressed in the direction of its length and bulges out at the sides, but it soon comes into a state of equilibrium, in which it continues to support the pressure; but if we substitute water for the india-rubber we cannot perform the experiment, for the water flows away laterally, and the jaws of the pincers come together without having exerted any appreciable pressure.

Bodies which can sustain a longitudinal pressure, however small that pressure may be, without being supported by a lateral pressure, are called solid bodies. Those which cannot do so are called fluids. We shall see that in a fluid at rest the pressure at any point must be equal in all directions, and this pressure is called the pressure of the fluid.

There are two great classes of fluids. If we put into a closed vessel a small quantity of a fluid of the first class, such as water, it will partly fill the vessel, and the rest of the vessel may either be empty or may contain a different fluid.

Fluids having this property are called liquids. Water is a liquid, and if we put a little water into a bottle the water will lie at the bottom of the bottle, and will be separated by a distinct surface from the air or the vapour of water above it.

If, on the contrary, the fluid which we put into the closed vessel be one of the second class, then, however small a portion we introduce, it will expand and fill the vessel, or at least as much of it as is not occupied by a liquid.

Fluids having this property are called gases. Air is a gas, and if we first exhaust the air from a vessel and then introduce the smallest quantity of air, the air will immediately expand till it fills the whole vessel so that there is as much air in a cubic inch in one part of the vessel as in another.

Hence a gas cannot, like a liquid, be kept in an open-mouthed vessel.

The distinction, therefore, between a gas and a liquid is that, however large the space may be into which a portion of gas is introduced, the gas will expand and exert pressure on every part of its boundary, whereas a liquid will not expand more than a very small fraction of its bulk, even when the pressure is reduced to zero; and some liquids can even sustain a hydrostatic tension, or negative pressure, without their parts being separated.

The three principal states in which bodies are found are, therefore, the solid, the liquid, and the gaseous states.

Most substances are capable of existing in all these states, as, for instance, water exists in the forms of ice, water, and steam. A few solids, such as carbon, have not yet been melted; and a few gases, such as oxygen, hydrogen, and nitrogen, have not yet been liquefied or solidified, but these may be considered as exceptional cases, arising from the limited range of temperature and pressure which we can command in our experiments.

The ordinary effects of heat in modifying the physical state of bodies may be thus described. We may take water as a familiar example, and explain, when it is necessary, the different phenomena of other bodies.

At the lowest temperatures at which it has been observed water exists in the solid form as ice. When heat is communicated to very cold ice, or to any other solid body not at its melting temperature—

1. The temperature rises.

2. The body generally expands (the only exception among solid bodies, as far as I am aware, is the iodide of silver, which has been found by M. Fizeau to contract as the temperature rises).

3. The rigidity of the body, or its resistance to change of form, generally diminishes. This phenomenon is more apparent in some bodies than in others. It is very conspicuous in iron, which when heated but not melted becomes soft and easily forged. The consistency of glass, resins, fats, and frozen oils alters very much with change of temperature. On the other hand, it is believed that steel wire is stiffer at 212° F. than at 32° F., and it has been shown by Joule and Thomson that the longitudinal elasticity of caoutchouc increases with the temperature between certain limits of temperature. When ice is very near its melting point it becomes very soft.

4. A great many solid bodies are constantly in a state of evaporation or transformation into the gaseous state at their free surface. Camphor, iodine, and carbonate of ammonia are well-known examples of this. These solid bodies, if not kept in stoppered bottles, gradually disappear by evaporation, and the vapour which escapes from them may be recognised by its smell and by its chemical action. Ice, too, is continually passing into a state of vapour at its surface, and in a dry climate during a long frost large pieces of ice become smaller and at last disappear.

There are other solid bodies which do not seem to lose any of their substance in this way; at least, we cannot detect any loss. It is probable, however, that those solid bodies which can be detected by their smell are evaporating with extreme slowness. Thus iron and copper have each a well-known smell. This, however, may arise from chemical action at the surface, which sets free hydrogen or some other gas combined with a very small quantity of the metal.

FUSION.

When the temperature of a solid body is raised to a sufficient height it begins to melt into a liquid. Suppose a small portion of the solid to be melted, and that no more heat is applied till the temperature of the remaining solid and of the liquid has become equalised; if a little more heat is then applied and the temperature again equalised there will be more liquid matter and less solid matter, but since the liquid and the solid are at the same temperature, that temperature must still be the melting temperature.

Hence, if the partly melted mass be kept well mixed together, so that the solid and fluid parts are at the same temperature, that temperature must be the melting temperature of the solid, and no rise of temperature will follow from the addition of heat till the whole of the solid has been converted into liquid.

The heat which is required to melt a certain quantity of a solid at the melting point into a liquid at the same temperature is called the latent heat of fusion.

It is called *latent* heat, because the application of this heat to the body does not raise its temperature or warm the body.

Those, therefore, who maintained heat to be a substance supposed that it existed in the fluid in a concealed or latent state, and in this way they distinguished it from the heat which, when applied to a body, makes it hotter, or raises the temperature. This they called sensible heat. A body, therefore, was said to possess so much heat. Part of this heat was called sensible heat, and to it was ascribed the temperature of the body. The other part was called latent heat, and to it was ascribed the liquid or gaseous form of the body.

The fact that a certain quantity of heat must be applied to a pound of melting ice to convert it into water is all that we mean in this treatise when we speak of this quantity of heat as the latent heat of fusion of a pound of water.

We make no assertion as to the state in which the heat exists in the water. We do not even assert that the heat communicated to the ice is still in existence as heat.

Besides the change from solid to liquid, there is generally a change of volume in the act of fusion. The water formed from the ice is of smaller bulk than the ice, as is shown by ice floating in water, so that the total volume of the ice and water diminishes as the melting goes on.

On the other hand, many substances expand in the act of fusion, so that the solid parts sink in the fluid. During the fusion of the mass the volume in these cases increases.

It has been shown by Prof. J. Thomson,[1] from the principles of the dynamical theory of heat, that if pressure is applied to a mixture of ice and water, it will not only compress both the ice and the water, but some of the ice will be melted at the same time, so that the total compression will be increased by the contraction of bulk due to this melting. The heat required to melt this ice being taken from the rest of the mass, the temperature of the whole will diminish.

Hence the melting point is lowered by pressure in the case of ice. This deduction from theory was experimentally verified by Sir W. Thomson.

If the substance had been one of those which expand in melting, the effect of pressure would be to solidify some of the mixture, and to raise the temperature of fusion. Most of the substances of which the crust of the earth is composed expand in the act of melting. Hence their melting points will rise under great pressure. If the earth were throughout

[1] *Transactions of the Royal Society of Edinburgh*, 1849.

in a state of fusion, when the external parts began to solidify they would sink in the molten mass, and when they had sunk to a great depth they would remain solid under the enormous pressure even at a temperature greatly above the point of fusion of the same rock at the surface. It does not follow, therefore, that in the interior of the earth the matter is in a liquid state, even if the temperature is far above that of the fusion of rocks in our furnaces.

It has been shown by Sir W. Thomson that if the earth, as a whole, were not more rigid than a ball of glass of equal size, the attraction of the moon and sun would pull it out of shape, and raise tides on the surface, so that the solid earth would rise and fall as the sea does, only not quite so much. It is true that this motion would be so smooth and regular that we should not be able to perceive it in a direct way, but its effect would be to diminish the apparent rise of the tides of the ocean, so as to make them much smaller than they actually are.

It appears, therefore, from what we know of the tides of the ocean, that the earth as a whole is more rigid than glass, and therefore that no very large portion of its interior can be liquid. The effect of pressure on the melting point of bodies enables us to reconcile this conclusion with the observed increase of temperature as we descend in the earth's crust, and the deductions as to the interior temperature founded on this fact by the aid of the theory of the conduction of heat.

EFFECT OF HEAT ON LIQUIDS.

When heat is applied to a liquid its effects are—

1. To warm the liquid. The quantity of heat required to raise the liquid one degree is generally greater than that required to raise the substance in the solid form one degree, and in general it requires more heat at high than at low temperatures to warm the liquid one degree.

2. To alter its volume. Most liquids expand as their

temperature rises, but water contracts from 32° F. to 39°·1 F., and then expands, slowly at first, but afterwards more rapidly.

3. To alter its physical state. Liquids, such as oil, tar, &c., which are sluggish in their motion, are said to be viscous. When they are heated their viscosity generally diminishes and they become more mobile. This is the case even with water, as appears by the experiments of M. O. E. Meyer.

When sulphur is heated, the melted sulphur undergoes several remarkable changes as its temperature rises, being mobile when first melted, then becoming remarkably viscous at a higher temperature, and again becoming mobile when still more heated. These changes are connected with chemical changes in the constitution of the sulphur, which are observed also in solid sulphur.

4. To convert the liquid into vapour.

Whatever be the temperature of a liquid which partially fills a vessel, it always gives off vapour till the remainder of the vessel is filled with vapour, and this goes on till the density of the vapour has reached a value which depends only on the temperature.

If in any way, as by the motion of a piston, the vessel be made larger, then more vapour will be formed till the density is the same as before. If the piston be pushed in, and the vessel made smaller, some of the vapour is condensed into the liquid state, but the density of the remainder of the vapour still remains the same.

If the remainder of the vessel, instead of containing nothing but the vapour of the liquid, contains any quantity of air or some other gas not capable of chemical action on the liquid, then exactly the same quantity of vapour will be formed, but the time required for the vapour to reach the further parts of the vessel will be greater, as it has to diffuse itself through the air in the vessel by a kind of percolation.

These laws of evaporation were discovered by Dalton.

The conversion of the liquid into vapour requires an amount of 'latent heat' which is generally much greater than the latent heat of fusion of the same substance.

In all substances, the density, pressure, and temperature are so connected that if we know any two of them the value of the third is determinate. Now in the case of vapours in contact with their own liquids or solids, there is for each temperature a corresponding density, which is the greatest density which the vapour can have at that temperature, without being condensed into the liquid or solid form.

Hence for each temperature there is also a maximum pressure which the vapour can exert.

A vapour which is at the maximum density and pressure corresponding to its temperature is called a *saturated* vapour. It is then just at the point of condensation, and the slightest increase of pressure or decrease of temperature will cause some of the vapour to be condensed. Professor Rankine restricts the use of the word vapour by itself to the case of a saturated vapour, and when the vapour is not at the point of condensation he calls it superheated vapour, or simply gas.

BOILING.

When a liquid in an open vessel is heated to a temperature such that the pressure of its vapour at that temperature is greater than the pressure at a point in the interior of the liquid, the liquid will begin to evaporate at that point, so that a bubble of vapour will be formed there. This process, in which bubbles of vapour are formed in the interior of the liquid, is called boiling or ebullition.

When water is heated in the ordinary way by applying heat to the bottom of a vessel, the lowest layer of the water becomes hot first, and by its expansion it becomes lighter than the colder water above, and gradually rises, so that a gentle circulation of water is kept up, and the whole water is gradually warmed, though the lowest layer is always the hottest. As the temperature increases, the absorbed air, which is generally found in ordinary water, is expelled, and rises in small bubbles without noise. At last the water in contact with the heated metal becomes so hot that, in spite of the pressure of the atmosphere on the surface of the water, the additional pressure due to the water in the vessel, and the cohesion of the water itself, some of the water at the bottom is transformed into steam, forming a bubble adhering to the bottom of the vessel. As soon as a bubble is formed, evaporation goes on rapidly from the water all round it, so that it soon grows large, and rises from the bottom. If the upper part of the water into which the bubble rises is still below the boiling temperature, the bubble is condensed, and its sides come together with a sharp rattling noise, called simmering. But the rise of the bubbles stirs the water about much more vigorously than the mere expansion of the water, so that the water is soon heated throughout, and brought to the boil, and then the bubbles enlarge rapidly during their whole ascent, and burst into the air, throwing the water about, and making the well-known softer and more rolling noise of boiling.

The steam, as it bursts out of the bubbles, is an invisible gas, but when it comes into the colder air it is cooled below its condensing point, and part of it is formed into a cloud consisting of small drops of water which float in the air. As the cloud of drops disperses itself and mixes with dry air the quantity of water in each cubic foot diminishes as the volume of any part of the cloud increases. The little drops of water begin to evaporate as soon as there is sufficient room for the vapour to be formed at the temperature of the atmosphere, and so the cloud vanishes again into thin air.

The temperature to which water must be heated before it boils depends, in the first place, on the pressure of the atmosphere, so that the greater the pressure, the higher the boiling temperature. But the temperature requires to be raised above that at which the pressure of steam is equal to that of the atmosphere, for the pressure of the vapour has to overcome not only the pressure due to the atmosphere and a certain depth of water, but that cohesion between the parts of the water of which the effects are visible in the tenacity of bubbles and drops. Hence it is possible to heat water 20° F. above its boiling point without ebullition. If a small quantity of metal-filings are now thrown into the water, the vapour by forming itself against the angular surface of these filings gets an advantage over the cohesion of the water, and produces a violent boiling, almost amounting to an explosion.

If a current of steam from a boiler is passed into a vessel of cold water, we have first the condensation of steam, accompanied with a very loud simmering noise, and a rapid heating of the water. When the water is sufficiently heated, the steam is not condensed, but escapes in bubbles, and the water is now boiling. If the boiler is at a high pressure, the steam from it will be at a temperature much above the boiling point in the open air, but in passing through the water in the open vessel it will cause some of it to evaporate, and when it issues from the water the temperature will be exactly that of the boiling point. For this reason, in finding the boiling point of a thermometer the instrument should not be allowed to dip in the water, but should be held in the steam.

As an instance of a different kind, let us suppose that the water is not pure, but contains some salt, such as common salt, or sulphate of soda, or any other substance which tends to combine with water, and from which the water must separate before it can evaporate. Water containing such substances in solution requires to be brought to a temperature higher than the boiling point of pure water before it will boil. Water, on the other hand, containing air or carbonic acid, will boil at a lower temperature than pure water till the gas is expelled.

If steam at 100° C. is passed into a vessel containing a

strong solution of one of the salts we have mentioned, which has a tendency to combine with water, the condensation of the steam will be promoted by this tendency, and will go on even after the solution has been heated far above the ordinary boiling point, so that by passing steam at 100° C. into a strong solution of nitrate of soda, Mr. Peter Spence[1] has heated it up to 121°·1 C.

The steam, however, which escapes, is still at 100° C.

If water at a temperature below 100° C. be placed in a vessel, and if by means of an air-pump we reduce the pressure of the air on the surface of the water, evaporation goes on and the surface of the water becomes colder than the interior parts. If we go on working the air-pump, the pressure is reduced to that of vapour of the temperature of the interior of the fluid. The water then begins to boil, exactly as in the ordinary way, and as it boils the temperature rapidly falls, the heat being expended in evaporating the water.

This experiment may be performed without an air-pump in the following way: Boil water in a flask over a gas-flame or spirit-lamp, and while it is boiling briskly cork the flask, and remove it from the flame. The boiling will soon cease, but if we now dash a little cold water over the flask, some of the steam in the upper part will be condensed, the pressure of the remainder will be diminished, and the water will begin to boil again. The experiment may be made more striking by plunging the flask entirely under cold water. The steam will be condensed as before, but the water, though it is cooled more rapidly than when the cold water was merely poured on the flask, retains its heat longer than the steam, and continues to boil for some time.

[1] *Transactions of the British Association*, 1869, p. 75.

ON THE GASEOUS STATE.

The distinguishing property of gases is their power of indefinite expansion. As the pressure is diminished the volume of the gas not only increases, but before the pressure has been reduced to zero the volume of the gas has become greater than that of any vessel we can put it in.

This is the property without which a substance cannot be called a gas, but it is found that actual gases fulfil with greater or less degrees of accuracy certain numerical laws, which are commonly referred to as the 'Gaseous Laws.'

LAW OF BOYLE.

The first of these laws expresses the relation between the pressure and the density of a gas, the temperature being constant, and is usually stated thus: 'The volume of a portion of gas varies inversely as the pressure.'

This law was discovered by Robert Boyle, and published by him in 1662, in an appendix to his 'New Experiments, Physico-mechanical, &c., touching the Spring of the Air.'

Mariotte, about 1676, in his treatise 'De la Nature de l'Air,' enunciated the same law, and carefully verified it, and it is generally referred to by Continental writers as Mariotte's law.

This law may also be stated thus:

The pressure of a gas is proportional to its density.

Another statement of the same law has been proposed by Professor Rankine, which I think places the law in a very clear light.

If we take a closed and exhausted vessel, and introduce into it one grain of air, this air will, as we know, exert a certain pressure on every square inch of the surface of the vessel. If we now introduce a second grain of air, then this second grain will exert exactly the same pressure on the sides of the vessel that it would have exerted if the first grain had not been there before it, so that the pressure will now be doubled. Hence we may state, as the property of a perfect gas, that any portion of it exerts the same pressure against the sides of a vessel as if the other portions had not been there.

Dalton extended this law to mixtures of gases of different kinds.

We have already seen that if several different portions of the same gas are placed together in a vessel, the pressure on any part of the sides of the vessel is the sum of the pressures which each portion would exert if placed by itself in the vessel.

Dalton's law asserts that the same is true for portions of different gases placed in the same vessel, and that the pressure of the mixture is the sum of the pressures due to the several portions of gas, if introduced separately into the vessel and brought to the same temperature.

This law of Dalton is sometimes stated as if portions of gas of different kinds behave to each other in a different manner from portions of gas of the same kind, and we are told that when gases of different kinds are placed in the same vessel, each acts as if the other were a vacuum.

This statement, properly understood, is correct, but it seems to convey the impression that if the gases had been of the same kind some other result would have happened, whereas there is no difference between the two cases.

Another law established by Dalton is that the maximum density of a vapour in contact with its liquid is not affected by the presence of other gases. It has been shown by M. Regnault that when the vapour of the substance has a tendency to combine with the gas, the maximum density attainable by the vapour is somewhat increased.

Before the time of Dalton it was supposed that the cause of evaporation was the tendency of water to combine with air, and that the water was dissolved in the air just as salt is dissolved in water.

Dalton showed that the vapour of water is a gas, which just at the surface of the water has a certain maximum density, and which will gradually diffuse itself through the space above, whether filled with air or not, till, if the space is limited, the density of the vapour is a maximum throughout, or, if the space is large enough, till the water is all dried up.

The presence of air is so far from being essential to this process that the more air there is, the slower it goes on, because the vapour has to penetrate through the air by the slow process of diffusion.

The phenomenon discovered by Regnault that the density of vapour is slightly increased by the presence of a gas which has a tendency to combine with it, is the only instance in which there is any truth in the doctrine of a liquid being held in solution by a gas.

The law of Boyle is not perfectly fulfilled by any actual gas. It is very nearly fulfilled by those gases which we are not able to condense into liquids, and among other gases it is most nearly fulfilled when their temperature is much above their point of condensation.

When a gas is near its point of condensation its density increases more rapidly than the pressure. When it is actually at the point of condensation the slightest increase of pressure condenses the whole of it into a liquid, and in the liquid form the density increases very slowly with the pressure.

LAW OF CHARLES.

The second law of gases was discovered by Charles,[1] but is commonly referred to as that of Gay-Lussac or of Dalton.[2] It may be stated thus :

[1] Professor of Physics at the Conservatoire des Arts et Métiers, Paris. Born 1746. Died 1823. Celebrated as having first employed hydrogen in balloons.

[2] Dalton, in 1801, first published this law. Gay-Lussac published it, in 1802, independently of Dalton. In his memoir, however (*Ann.*

The volume of a gas under constant pressure expands when raised from the freezing to the boiling temperature by the same fraction of itself, whatever be the nature of the gas.

It has been found by the careful experiments of M. Regnault, M. Rudberg, Prof. B. Stewart, and others that the volume of air at constant pressure expands from 1 to 1·3665 between 0° C. and 100° C. Hence 30 cubic inches of air at 0° C. would expand to about 41 cubic inches at 100° C.

If we admit the truth of Boyle's law at all temperatures, and if the law of Charles is found to be true for a particular pressure, say that of the atmosphere, then it is easy to show that the law of Charles must be true for every other pressure. For if we call the volume V and the pressure P, then we may call the product of the numerical value of the volume and pressure V P, and Boyle's law asserts that this product is constant, provided the temperature is constant. If then we are further informed that when P has a given value V is increased from 1 to 1·3665 when the temperature rises from the freezing point to the boiling point, the product V P will be increased in the same proportion at that particular pressure. But V P we know by Boyle's law does not depend on the particular pressure, but remains the same for all pressures when the temperature remains the same. Hence, whatever be the pressure, the product V P will be increased in the proportion of 1 to 1·3665 when the temperature rises from 0° C. to 100° C.

The law of the equality of the dilatation of gases, which, as originally stated, applied only to the dilatation from 0° C. to 100° C., has been found to be true for all other temperatures for which it has hitherto been tested.

de Chimie, xliii. p. 157 [1802]), he states that Citizen Charles had remarked, fifteen years before the date of his memoir, the equality of the dilatation of the principal gases; but, as Charles never published these results, he had become acquainted with them by mere chance.

It appears, therefore, that gases are distinguished from other forms of matter, not only by their power of indefinite expansion so as to fill any vessel, however large, and by the great effect which heat has in dilating them, but by the uniformity and simplicity of the laws which regulate these changes. In the solid and liquid states the effect of a given change of pressure or of temperature in changing the volume of the body is different for every different substance. On the other hand, if we take equal volumes of any two gases, measured at the same temperature and pressure, their volumes will remain equal if we afterwards bring them both to any other temperature and pressure, and this although the two gases differ altogether in chemical nature and in density, provided they are both in the perfectly gaseous condition.

This is only one of many remarkable properties which point out the gaseous state of matter as that in which its physical properties are least complicated.

In our description of the physical properties of bodies as related to heat we have begun with solid bodies, as those which we can most easily handle, and have gone on to liquids, which we can keep in open vessels, and have now come to gases, which will escape from open vessels, and which are generally invisible. This is the order which is most natural in our first study of these different states. But as soon as we have been made familiar with the most prominent features of these different conditions of matter, the most scientific course of study is in the reverse order, beginning with gases, on account of the greater simplicity of their laws, then advancing to liquids, the more complex laws of which are much more imperfectly known, and concluding with the little that has been hitherto discovered about the constitution of solid bodies.

CHAPTER II.

ON THERMOMETRY, OR THE THEORY OF TEMPERATURE.

Definition of Temperature.—*The temperature of a body is its thermal state considered with reference to its power of communicating heat to other bodies.*

Definition of Higher and Lower Temperature.—*If when two bodies are placed in thermal communication, one of the bodies loses heat, and the other gains heat, that body which gives out heat is said to have a higher temperature than that which receives heat from it.*

Cor. *If when two bodies are placed in thermal communication neither of them loses or gains heat, the two bodies are said to have equal temperatures or the same temperature. The two bodies are then said to be in thermal equilibrium.* We have here a means of comparing the temperature of any two bodies, so as to determine which has the higher temperature, and a test of the equality of temperature which is independent of the nature of the bodies tested. But we have no means of estimating numerically the difference between two temperatures, so as to be able to assert that a certain temperature is exactly halfway between two other temperatures.

Law of Equal Temperatures.—*Bodies whose temperatures are equal to that of the same body have themselves equal temperatures.* This law is not a truism, but expresses the fact that if a piece of iron when plunged into a vessel of water is in thermal equilibrium with the water, and if the same piece of iron, without altering its temperature, is transferred to a vessel of oil, and is found to be also in thermal equilibrium with the oil, then if the oil and water were put into the same vessel they would themselves be in thermal equilibrium, and the same would be true of any other three substances.

This law, therefore, expresses much more than Euclid's axiom that 'Things which are equal to the same thing are equal to one another,' and is the foundation of the whole science of thermometry. For if we take a thermometer, such as we have already described, and bring it into intimate contact with different bodies, by plunging it into liquids, or inserting it into holes made in solid bodies, we find that the mercury in the tube rises or falls till it has reached a certain point at which it remains stationary. We then know that the thermometer is neither becoming hotter nor colder, but is in thermal equilibrium with the surrounding body. It follows from this, by the law of equal temperatures, that the temperature of the body is the same as that of the thermometer, and the temperature of the thermometer itself is known from the height at which the mercury stands in the tube.

Hence the *reading*, as it is called, of the thermometer—that is, the number of degrees indicated on the scale by the top of the mercury in the tube—informs us of the temperature of the surrounding substance, as well as of that of the mercury in the thermometer. In this way the thermometer may be used to compare the temperature of any two bodies at the same time or at different times, so as to ascertain whether the temperature of one of them is higher or lower than that of the other. We may compare in this way the temperatures of the air on different days; we may ascertain that water boils at a lower temperature at the top of a mountain than it does at the sea-shore, and that ice melts at the same temperature in all parts of the world.

For this purpose it would be necessary to carry the same thermometer to different places, and to preserve it with great care, for if it were destroyed and a new one made, we should have no certainty that the same temperature is indicated by the same reading in the two thermometers.

Thus the observations of temperature recorded during sixteen years by Rinieri[1] at Florence lost their scientific value after the suppression of the Accademia del Cimento, and the supposed destruction of the thermometers with which the observations were made. But when Antinori in 1829 discovered a number of the very thermometers used in the ancient observations, Libri[2] was able to compare them with Réaumur's scale, and thus to show that the climate of Florence has not been rendered sensibly colder in winter by the clearing of the woods of the Apennines.

In the construction of artificial standards for the measurement of quantities of any kind it is desirable to have the means of comparing the standards together, either directly, or by means of some natural object or phenomenon which is easily accessible and not liable to change. Both methods are used in the preparation of thermometers.

We have already noticed two natural phenomena which take place at definite temperatures—the melting of ice and the boiling of water. The advantage of employing these temperatures to determine two points on the scale of the thermometer was pointed out by Sir Isaac Newton ('Scala Graduum Caloris,' Phil. Trans. 1701).

The first of these points of reference is commonly called the Freezing Point. To determine it, the thermometer is placed in a vessel filled with pounded ice or snow thoroughly moistened with water. If the atmospheric temperature be above the freezing point, the melting of the ice will ensure the presence of water in the vessel. As long as every part of the vessel contains a mixture of water and ice its temperature remains uniform, for if heat enters the vessel it can only melt some of the ice, and if heat escapes from the vessel some of the water will freeze, but the mixture can be made neither hotter nor colder till all the ice is melted or all the water frozen.

[1] Pupil of Galileo; died 1647.

[2] *Annales de Chimie et de Physique*, xlv. (1830).

The thermometer is completely immersed in the mixture of ice and water for a sufficient time, so that the mercury has time to reach its stationary point. The position of the top of the mercury in the tube is then recorded by making a scratch on the glass tube. We shall call this mark the Freezing Point. It may be determined in this way with extreme accuracy, for, as we shall see afterwards, the temperature of melting ice is very nearly the same under very different pressures.

Fig. 2.

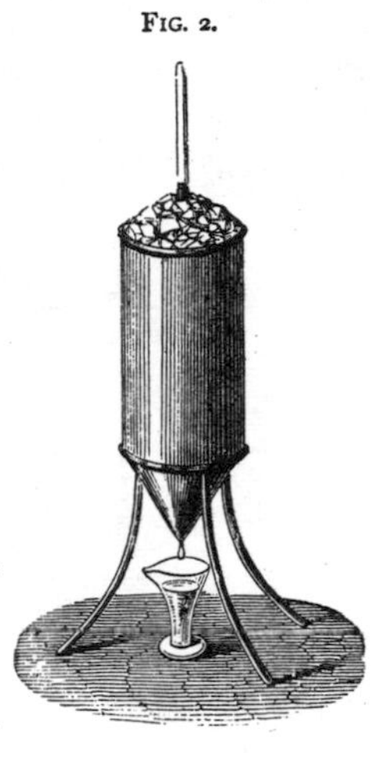

The other point of reference is called the Boiling Point. The temperature at which water boils depends on the pressure of the atmosphere. The greater the pressure of the air on the surface of the water, the higher is the temperature to which the water must be raised before it begins to boil.

To determine the Boiling Point, the stem of the thermometer is passed through a hole in the lid of a tall vessel, in the lower part of which water is made to boil briskly, so that the whole of the upper part, where the thermometer is placed, is filled with steam. When the thermometer has acquired the temperature of the current of steam the stem is drawn up through the hole in the lid of the vessel till the top of the column of mercury becomes visible. A scratch is then made on the tube to indicate the boiling point.

In careful determinations of the boiling point no part of the thermometer is allowed to dip into the boiling water, because it has been found by Gay-Lussac that the temperature of the water is not always the same, but that it boils at different temperatures in different kinds of vessels. It has been shown, however, by Rudberg that the temperature of the steam which escapes from boiling water is the same in every kind of vessel, and depends only on the pressure at the surface of the water. Hence the thermometer is not dipped in the water, but suspended in the issuing steam. To ensure that the temperature of the steam shall be the same when it reaches the thermometer as when it issues from the boiling water, the sides of the vessel are sometimes protected by what is called a steam-jacket. A current of steam is made to play over the outside of the sides of the vessel. The vessel is thus raised to the same temperature as the steam itself, so that the steam cannot be cooled during its passage from the boiling water to the thermometer.

FIG. 3.

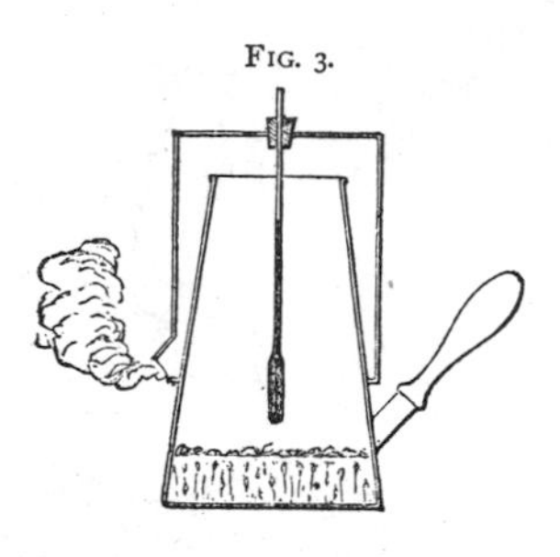

For instance, if we take any tall narrow vessel, as a coffee-pot, and cover its mouth and part of its sides with a wider vessel turned upside down, taking care that there shall be plenty of room for the steam to escape, then if we boil a small quantity of water in the coffee-pot, a thermometer placed in the steam above will be raised to the exact temperature of the boiling point of water corresponding to the state of the barometer at the time.

To mark the level of the mercury on the tube of the thermometer without cooling it, we must draw it up through a cork or a plug of india-rubber in the steam-jacket through which the steam passes till we can just see the top of the column of mercury. A mark must then be scratched on the glass to register the boiling point. This experiment of exposing a thermometer to the steam of boiling water is an important one, for it not only supplies a means of graduating thermometers, and testing them when they have been graduated, but, since the temperature at which water boils depends on the pressure of the air, we may determine the pressure of the air by boiling water when we are not able to measure it by means of the appropriate instrument, the barometer.

We have now obtained two points of reference marked by scratches on the tube of the thermometer—the freezing point and the boiling point. We shall suppose for the present that when the boiling point was marked the barometer happened to indicate the standard pressure of 29·905 inches of mercury at 32° F. at the level of the sea in the latitude of London. In this case the boiling point is the standard boiling point. In any other case it must be corrected.

Our thermometer will now agree with any other properly constructed thermometer at these two temperatures.

In order to indicate other temperatures, we must construct a scale—that is, a series of marks—either on the tube itself or on a convenient part of the apparatus close to the tube and well fastened to it.

For this purpose, having settled what values we are to give to the freezing and the boiling points, we divide the space between those points into as many equal parts as there are degrees between them, and continue the series of equal divisions up and down the scale as far as the tube of the thermometer extends.

Three different ways of doing this are still in use, and, as we often find temperatures stated according to a different scale from that which we adopt ourselves, it is necessary to know the principles on which these scales are formed.

The Centigrade scale was introduced by Celsius.[1] In it the freezing point is marked 0° and called zero, and the boiling point is marked 100°.

The obvious simplicity of this mode of dividing the space between the points of reference into 100 equal parts and

[1] Professor of Astronomy in the University of Upsala.

calling each of these a degree, and reckoning all temperatures in degrees from the freezing point, caused it to be very generally adopted, along with the French decimal system of measurement, by scientific men, especially on the Continent of Europe. It is true that the advantage of the decimal system is not so great in the measurement of temperatures as in other cases, as it merely makes it easier to remember the freezing and boiling temperatures, but the graduation is not too fine for the roughest purposes, while for accurate measurements the degrees may be subdivided into tenths and hundredths.

The other two scales are called by the names of those who introduced them.

Fahrenheit, of Dantzig, about 1714, first constructed thermometers comparable with each other. In Fahrenheit's scale the freezing point is marked 32°, and the boiling point 212°, the space between being divided into 180 equal parts, and the graduation extended above and below the points of reference. A point 32 degrees below the freezing point is called zero, or 0°, and temperatures below this are indicated by the number of degrees below zero.

This scale is very generally used in English-speaking countries for purposes of ordinary life, and also for those of science, though the Centigrade scale is coming into use among those who wish their results to be readily followed by foreigners.

The only advantages which can be ascribed to Fahrenheit's scale, besides its early introduction, its general diffusion, and its actual employment by so many of our countrymen, are that mercury expands almost exactly one ten-thousandth of its volume at 142° F. for every degree of Fahrenheit's scale, and that the coldest temperature which we can get by mixing snow and salt is near the zero of Fahrenheit's scale.

To compare temperatures given in Fahrenheit's scale with temperatures given in the Centigrade scale we have only to remember that 0° Centigrade is 32° Fahrenheit, and that five degrees Centigrade are equal to nine of Fahrenheit.

The third thermometric scale is that of Réaumur. In this scale the freezing point is marked 0° and the boiling point 80°. I am not aware of any advantage of this scale. It is used to some extent on the Continent of Europe for medical and domestic purposes. Four degrees of Réaumur correspond to five Centigrade and to nine of Fahrenheit.

The existence of these three thermometric scales furnishes an example of the inconvenience of the want of uniformity in systems of measurement. The whole of what we have said about the comparison of the different scales might have been omitted if any one of these scales had been adopted by all who use thermometers. Instead of spending our time in describing the arbitrary proposals of different men, we should have gone on to investigate the laws of heat and the properties of bodies.

We shall afterwards have occasion to use a scale differing in its zero-point from any of those we have considered, but when we do so we shall bring forward reasons for its adoption depending on the nature of things and not on the predilections of men.

If two thermometers are constructed of the same kind of glass, with tubes of uniform bore, and are filled with the same liquid and then graduated in the same way, they may be considered for ordinary purposes as comparable instruments; so that though they may never have been actually compared together, yet in ascertaining the temperature of anything there will be very little difference whether we use the one thermometer or the other.

But if we desire great accuracy in the measurement of temperature, so that the observations made by different observers with different instruments may be strictly comparable, the only satisfactory method is by agreeing to choose one thermometer as a standard and comparing all the others with it.

All thermometers ought to be made with tubes of as uniform bore as can be found; but for a standard thermometer the bore should be calibrated—that is to say, its size should be measured at short intervals all along its length.

For this purpose, before the bulb is blown, a small quantity of mercury is introduced into the tube and moved along the tube by forcing air into the tube behind it. This is done by squeezing the air out of a small india-rubber ball which is fastened to the end of the tube.

If the length of the column of mercury remains exactly the same as it passes along the tube, the bore of the tube must be uniform; but even in the best tubes there is always some want of uniformity.

But if we introduce a short column of mercury into the tube, then mark both ends of the column, and move it on its own length, till one end comes exactly to the mark where the other end was originally, then mark the other end, and move it on again, we shall have a series of marks on the tube such that the capacity of the tube between any two consecutive marks will be the same, being equal to that of the column of mercury.

By this method, which was invented by Gay-Lussac, a number of divisions may be marked on the tube, each of which contains the same volume, and though they will probably not correspond to degrees when the tube is made up into a thermometer, it will be easy to convert the reading of this instrument into degrees by multiplying it by a proper factor, and in the use of a standard instrument this trouble is readily undertaken for the sake of accuracy.

The tube having been prepared in this way, one end is heated till it is melted, and it is blown into a bulb by forcing air in at the other end of the tube. In order to avoid introducing moisture into the tube, this is done, not by the mouth, but by means of a hollow india-rubber ball, which is fastened to the end of the tube.

The tube of a thermometer is generally so narrow that mercury will not enter it, for a reason which we shall explain when we come to the properties of liquids. Hence the following method is adopted to fill the thermometer. By rolling paper round the open end of the tube, and making the tube thus formed project a little beyond the glass tube, a cavity is formed, into which a little mercury is poured. The mercury, however, will not run down the tube of the thermometer, partly because the bulb and tube are already full of air, and partly because the mercury requires a certain pressure from without to enter so narrow a tube. The bulb is therefore gently heated so as to cause the air to expand, and some of the air escapes through the mercury. When the bulb cools, the pressure of the air in the bulb becomes less than the pressure of the air outside, and the difference of these pressures is sufficient to make the mercury enter the tube, when it runs down and partially fills the bulb.

Fig. 4.

In order to get rid of the remainder of the air, and of any moisture in the thermometer, the bulb is gradually heated till the mercury boils. The air and steam escape along with the vapour of mercury, and as the boiling continues the last remains of air are expelled through the mercury at the top of the tube. When the boiling ceases, the mercury runs back into the tube, which is thus perfectly filled with mercury.

While the thermometer is still hotter than any temperature at which it will afterwards be used, and while the mercury or its vapour completely fills it, a blowpipe flame is made to play on the top of the tube, so as to melt it and close the end of the tube. The tube, thus closed with its own substance, is said to be 'hermetically sealed.'[1]

There is now nothing in the tube but mercury, and when the mercury contracts so as to leave a space above it, this space is either empty of all gross matter, or contains only the vapour of mercury. If, in spite of all our precautions, there is still some air in the tube, this can easily be ascertained by inverting the thermometer and letting some of the mercury glide towards the end of the tube. If the instrument is perfect, it will reach the end of the tube and completely fill it. If there is air in the tube the air will form an elastic cushion, which will prevent the mercury from reaching the end of the tube, and will be seen in the form of a small bubble.

We have next to determine the freezing and boiling points, as has been already described, but certain precautions have still to be observed. In the first place, glass is a substance in which internal changes go on for some time after it has been strongly heated, or exposed to intense forces. In fact, glass is in some degree a plastic body. It is found that after a thermometer has been filled and sealed the capacity of the bulb diminishes slightly, and that this change is comparatively rapid at first, and only gradually becomes insensible as the bulb approaches its ultimate condition. It causes the freezing point to rise in the tube to $0^{\circ}\cdot3$ or $0^{\circ}\cdot5$, and if, after the displacement of the zero, the mercury be again boiled, the zero returns to its old place and gradually rises again.

This change of the zero-point was discovered by M. Flaugergues.[2] It may be considered complete in from four to

[1] 'From Hermes or Mercury, the imagined inventor of chemistry.'—*Johnson's Dict.*

[2] *Ann. de Chimie et de Physique*, xxi. p. 333 (1822).

six months.[1] In order to avoid the error which it would introduce into the scale, the instrument should, if possible, have its zero determined some months after it has been filled, and since even the determination of the boiling point of water produces a slight depression of the freezing point (that is, an expansion of the bulb), the freezing point should not be determined after the boiling point, but rather before it.

When the boiling point is determined, the barometer is probably not at the standard height. The mark made on the thermometer must, in graduating it, be considered to represent, not the standard boiling point, but the boiling point corresponding to the observed height of the barometer, which may be found from the tables.

To construct a thermometer in this elaborate way is by no means an easy task, and even when two thermometers have been constructed with the utmost care, their readings at points distant from the freezing and boiling points may not agree, on account of differences in the law of expansion of the glass of the two thermometers. These differences, however, are small, for all thermometers are made of the same description of glass.

But since the main object of thermometry is that all thermometers shall be strictly comparable, and since thermometers are easily carried from one place to another, the best method of obtaining this object is by comparing all thermometers either directly or indirectly with a single standard thermometer. For this purpose, the thermometers, after being properly graduated, are all placed along with the standard thermometer in a vessel, the temperature of which can be maintained uniform for a considerable time. Each thermometer is then compared with the standard thermometer.

[1] Dr. Joule, however, finds that the rise of the freezing point of a delicate thermometer has been going on for twenty-six years, though the changes are now exceedingly minute.—*Phil. Soc. Manchester*, Feb. 22, 1870.

A table of corrections is made for each thermometer by entering the reading of that thermometer, along with the correction which must be applied to that reading to reduce it to the reading of the standard thermometer. This is called the proper correction for that reading. If it is positive it must be added to the reading, and if negative it must be subtracted from it.

By bringing the vessel to various temperatures, the corrections at these temperatures for each thermometer are ascertained, and the series of corrections belonging to each thermometer is made out and preserved along with that thermometer.

Any thermometer may be sent to the Observatory at Kew, and will be returned with a list of corrections, by the application of which, observations made with that thermometer become strictly comparable with those made by the standard thermometer at Kew, or with any other thermometer similarly corrected. The charge for making the comparison is very small compared with the expense of making an original standard thermometer, and the scientific value of observations made with a thermometer thus compared is greater than that of observations made with the most elaborately prepared thermometer which has not been compared with some existing and known standard instrument.

I have described at considerable length the processes by which the thermometric scale is constructed, and those by which copies of it are multiplied, because the practical establishment of such a scale is an admirable instance of the method by which we must proceed in the scientific observation of a phenomenon such as temperature, which, for the present, we regard rather as a *quality*, capable of greater or less intensity, than as a *quantity* which may be added to or subtracted from other quantities of the same kind.

A temperature. so far as we have yet gone in the science of heat, is not considered as capable of being added to another temperature so as to form a temperature which is the sum of its components. When we are able to attach a distinct meaning to such an operation, and determine its result, our conception of temperature will be raised to the rank of a quantity. For the present, however, we must be content to regard temperature as a quality of bodies, and be satisfied to know that the temperatures of all bodies can be referred to their proper places in the same scale.

For instance, we have a right to say that the temperatures of freezing and boiling differ by 180° Fahrenheit; but we have as yet no right to say that this difference is the same as that between the temperatures 300° and 480° on the same scale. Still less can we assert that a temperature of

$$244^\circ \text{ F.} = 32^\circ + 212^\circ$$

is equal to the *sum* of the temperatures of freezing and boiling. In the same way, if we had nothing by which to measure time except the succession of our own thoughts, we might be able to refer each event within our own experience to its proper chronological place in a series, but we should have no means of comparing the interval of time between one pair of events with that between another pair, unless it happened that one of these pairs was included within the other pair, in which case the interval between the first pair must be the smallest. It is only by observation of the uniform or periodic motions of bodies, and by ascertaining the conditions under which certain motions are always accomplished in the same time, that we have been enabled to measure time, first by days and years, as indicated by the heavenly motions, and then by hours, minutes, and seconds, as indicated by the pendulums of our clocks, till we are now able, not only to calculate the time of vibration of different kinds of light, but to compare the time of vibration of a molecule of hydrogen set in motion by an electric discharge through a glass tube, with the time of vibration of another molecule of hydrogen in the sun, forming part of some great eruption of rosy clouds, and with the time of vibration of another molecule in Sirius which has not transmitted its vibrations to our earth, but has simply prevented vibrations arising in the body of that star from reaching us.

In a subsequent chapter we shall consider the further progress of our knowledge of Temperature as a Quantity.

ON THE AIR THERMOMETER.

The original thermometer invented by Galileo was an air thermometer. It consisted of a glass bulb with a long neck. The air in the bulb was heated, and then the neck was plunged into a coloured liquid. As the air in the bulb cooled, the liquid rose in the neck, and the higher the liquid the lower the temperature of the air in the bulb. By putting the bulb into the mouth of a patient, and noting the point to which the liquid was driven down in the tube, a physician might estimate whether the ailment was of the nature of a fever or not. Such a thermometer has several obvious merits. It is easily constructed, and gives larger indications for the same change of temperature than a thermometer containing any liquid as the thermometric substance. Besides this, the air requires less heat to warm it than an equal bulk of any liquid, so that the air thermometer is very rapid in its indications. The great inconvenience of the instrument as a means of measuring temperature is, that the height of the liquid in the tube depends on the pressure of the atmosphere as well as on the temperature of the air in the bulb. The air thermometer cannot therefore of itself tell us anything about temperature. We must consult the barometer at the same time, in order to correct the reading of the air thermometer. Hence the air thermometer, to be of any scientific value, must be used along with the barometer, and its readings are of no use till after a process of calculation has been gone through. This puts it at a great disadvantage compared with the mercurial thermometer as a means of ascertaining temperatures. But if the researches on which we are engaged are of so important a nature that we are willing to undergo the labour of double observations and numerous calculations, then the advantages of the air thermometer may again preponderate.

We have seen that in fixing a scale of temperature after marking on our thermometer two temperatures of reference and filling up the interval with equal divisions, two thermometers containing different liquids will not in general agree except at the temperatures of reference.

If, on the other hand, we could secure a constant pressure in the air thermometer, then if we exchange the air for any other gas, all the readings will be exactly the same provided the reading at one of the temperatures of reference is the same. It appears, therefore, that the scale of temperatures as indicated by an air thermometer has this advantage over the scale indicated by mercury or any other liquid or solid, that whereas no two liquid or solid substances can be made to agree in their expansion throughout the scale, all the gases agree with one another. In the absence of any better reasons for choosing a scale, the agreement of so many substances is a reason why the scale of temperatures furnished by the expansion of gases should be considered as of great scientific value. In the course of our study we shall find that there are scientific reasons of a much higher order which enable us to fix on a scale of temperature, based not on a probability of this kind, but on a more intimate knowledge of the properties of heat. This scale, so far as it has been investigated, is found to agree very closely with that of the air thermometer.

There is another reason, of a practical kind, in favour of the use of air as a thermometric substance, namely, that air remains in the gaseous state at the lowest as well as the highest temperatures which we can produce, and there are no indications in either case of its approaching to a change of state. Hence air, or one of the permanent gases, is of the greatest use in estimating temperatures lying far outside of the temperatures of reference, such, for instance, as the freezing point of mercury or the melting point of silver.

We shall consider the practical method of using air as a thermometric substance when we come to Gasometry. In the meantime let us consider the air thermometer in its simplest form, that of a long tube of uniform bore closed at one end, and containing air or some other gas which is separated from the outer air by a short column of mercury, oil, or some other liquid which is capable of moving freely along the tube, while at the same time it prevents all communication between the confined air and the atmosphere. We shall also suppose that the pressure acting on the confined air is in some way maintained constant during the course of the experiments we are going to describe.

Fig. 5.

Air Thermometer.

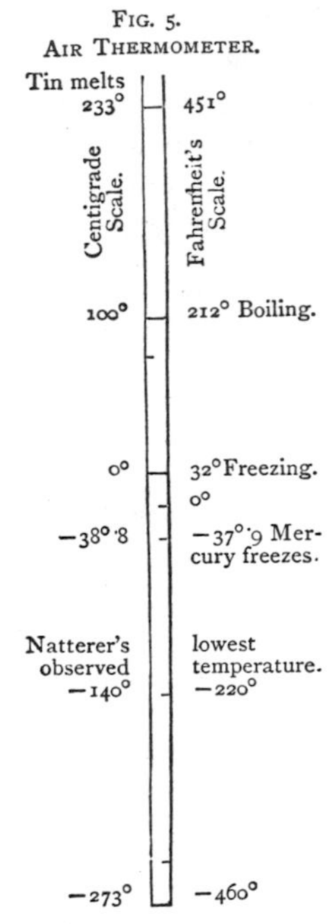

The air thermometer is first surrounded with ice and ice-cold water. Let us suppose that the upper surface of the air now stands at the point marked 'Freezing.' The thermometer is then surrounded with the steam rising from water boiling under an atmospheric pressure of 29·905 inches of mercury. Let the surface of the enclosed air now stand at the point marked 'Boiling.' In this way, the two temperatures of reference are to be marked on the tube.

To complete the scale of the thermometer we must divide the distance between boiling and freezing into a selected number of equal parts, and carry this graduation up and down the tube beyond the freezing and boiling points with degrees of the same length.

Of course, if we carry the graduation far enough down the tube, we shall at last come to the bottom of the tube. What will be the reading at that point? and what is meant by it?

To determine the reading at the bottom of the tube is a very simple matter. We know that the distance of the freezing point from the bottom of the tube is to the distance of the boiling point from the bottom in the proportion of 1 to 1·3665, since this is the dilatation of air between the freezing and the boiling temperatures. Hence it follows, by an easy arithmetical calculation, that if, as in Fahrenheit's scale, the freezing point is marked 32°, and the boiling point 212°, the bottom of the tube must be marked −459°·13. If, as in the Centigrade scale, the freezing point is marked 0°, and the boiling point 100°, the bottom of the tube will be marked −272°·85. This, then, is the reading at the bottom of the scale.

The other question, What is meant by this reading requires a more careful consideration. We have begun by defining the measure of the temperature as the reading of the scale of our thermometer when it is exposed to that temperature. Now if the reading could be observed at the bottom of the tube, it would imply that the volume of the air had been reduced to nothing. It is hardly necessary to say that we have no expectation of ever observing such a reading. If it were possible to abstract from a substance all the heat it contains, it would probably still remain an extended substance, and would occupy a certain volume Such an abstraction of all its heat from a body has never been effected, so that we know nothing about the temperature which would be indicated by an air thermometer placed in contact with a body absolutely devoid of heat. This much we are sure of, however, that the reading would be above −459°·13 F.

It is exceedingly convenient, especially in dealing with questions relating to gases, to reckon temperatures, not from the freezing point, or from Fahrenheit's zero, but from the bottom of the tube of the air thermometer.

This point is then called the absolute zero of the air thermometer, and temperatures reckoned from it are called absolute temperatures. It is probable that the dilatation of a perfect gas is a little less than 1·3665. If we suppose it 1·366, then absolute zero will be $-460°$ on Fahrenheit's scale, or $-273°\frac{1}{3}$ Centigrade.

If we add 460° to the ordinary reading on Fahrenheit's scale, we shall obtain the absolute temperature in Fahrenheit's degrees.

If we add $273°\frac{1}{3}$ to the Centigrade reading, we shall obtain the absolute temperature in Centigrade degrees.

We shall often have occasion to speak of absolute temperature by the air thermometer. When we do so we mean nothing more than what we have just said—namely, temperature reckoned from the bottom of the tube of the air thermometer. We assert nothing as to the state of a body deprived of all its heat, about which we have no experimental knowledge.

One of the most important applications of the conception of absolute temperature is to simplify the expression of the two laws discovered respectively by Boyle and by Charles. The laws may be combined into the statement that *the product of the volume and pressure of any gas is proportional to the absolute temperature.*

For instance, if we have to measure quantities of a gas by their volumes under various conditions as to temperature and pressure, we can reduce these volumes to what they would be at some standard temperature and pressure.

Thus if V, P, T be the actual volume, pressure, and absolute temperature, and V_0 the volume at the standard pressure P_0, and the standard temperature T_0, then

$$\frac{V\,P}{T} = \frac{V_0\,P_0}{T_0}$$

or

$$V_0 = V\frac{P}{P_0}\frac{T_0}{T}$$

If we have only to compare the relative quantities of the gas in different measurements in the same series of experiments, we may suppose P_0 and T_0 both unity, and use the quantity $\frac{VP}{T}$ without always multiplying it by $\frac{T_0}{P_0}$, which is a constant quantity.[1]

The great scientific importance of the scale of temperature as determined by means of the air or gas thermometer arises from the fact, established by the experiments of Joule and Thomson, that the scale of temperature derived from the expansion of the more permanent gases is almost exactly the same as that founded upon purely thermodynamic considerations, which are independent of the peculiar properties of the thermometric body. This agreement has been experimentally verified only within a range of temperature between 0° C. and 100° C. If, however, we accept the molecular theory of gases, the volume of a perfect gas ought to be exactly proportional to the absolute temperature on the thermodynamic scale, and it is probable that as the temperature rises the properties of real gases approximate to those of the theoretically perfect gas.

All the thermometers which we have considered have been constructed on the principle of measuring the expansion of a substance as the temperature rises. In certain cases it is convenient to estimate the temperature of a substance by the heat which it gives out as it cools to a standard temperature. Thus if a piece of platinum heated in a furnace is dropped into water, we may form an estimate of the temperature of the furnace by the amount of heat communicated to the water. Some have supposed that this method of estimating temperatures is more scientific than that founded on expansion. It would be so if the same quantity of heat always caused the same rise of temperature, whatever the original

[1] For a full account of the methods of measuring gases the student is referred to Bunsen's *Gasometry*, translated by Roscoe.

temperature of the body. But the specific heat of most substances increases as the temperature rises, and it increases in different degrees for different substances, so that this method cannot furnish an absolute scale of temperature. It is only in the case of gases that the specific heat of a given mass of the substance remains the same at all temperatures.

There are two methods of estimating temperature which are founded on the electrical properties of bodies. We cannot, within the limits of this treatise, enter into the theory of these methods, but must refer the student to works on electricity. One of these methods depends on the fact that in a conducting circuit formed of two different metals, if one of the junctions be warmer than the other, there will be an electromotive force which will produce a current of electricity in the circuit, and this may be measured by means of a galvanometer. In this way very minute differences of temperature between the ends of a piece of metal may be detected. Thus if a piece of iron wire is soldered at both ends to a copper wire, and if one of the junctions is at a place where we cannot introduce an ordinary thermometer, we may ascertain its temperature by placing the other junction in a vessel of water and adjusting the temperature of the water till no current passes. The temperature of the water will then be equal to that of the inaccessible junction.

Electric currents excited by differences of temperature in different parts of a metallic circuit are called thermo-electric currents. An arrangement by which the electromotive forces arising from a number of junctions may be added together is called a thermopile, and is used in experiments on the heating effect of radiation, because it is more sensitive to changes of temperature caused by small quantities of heat than any other instrument.

Professor Tait[1] has found that if t_1 and t_2 denote the temperatures of the hot and cold junction of two metals,

[1] *Proceedings of the Royal Society of Edinburgh*, 1870-71.

the electromotive force of the circuit formed by these two metals is

$$\text{A}\,(t_1 - t)\left[\text{T} - \tfrac{1}{2}(t_1 + t_2)\right],$$

where A is a constant depending on the nature of the metals, and T is a temperature also depending on the metals, such that when one junction is as much hotter than T as the other is colder, no current is produced. T may be called the neutral temperature for the two metals. For copper and iron it is about 284° C.

The other method of estimating the temperature of a place at which we cannot set a thermometer is founded on the increase of the electric resistance of metals as the temperature rises. This method has been successfully employed by Mr. Siemens.[1] Two coils of the same kind of fine platinum wire are prepared so as to have equal resistance. Their ends are connected with long thick copper wires, so that the coils may be placed if necessary a long way from the galvanometer. These copper terminals are also adjusted so as to be of the same resistance for both coils. The resistance of the terminals should be small as compared with that of the coils. One of the coils is then sunk, say to the bottom of the sea, and the other is placed in a vessel of water, the temperature of which is adjusted till the resistance of both coils is the same. By ascertaining with a thermometer the temperature of the vessel of water, that of the bottom of the sea may be deduced.

Mr. Siemens has found that the resistance of the metals may be expressed by a formula of the form

$$\text{R} = \alpha\sqrt{\text{T}} + \beta\,\text{T} + \gamma,$$

where R is the resistance, T the absolute temperature, and α β γ coefficients. Of these α is the largest, and the resistance depending on it increases as the square root of the absolute temperature, so that the resistance increases more slowly as the temperature rises. The second term, β T, is

[1] *Proceedings of the Royal Society*, April 27, 1871.

proportional to the temperature and may be attributed to the expansion of the substance. The third term is constant.

CHAPTER III.

CALORIMETRY.

HAVING explained the principles of Thermometry, or the method of ascertaining temperatures, we are able to understand what we may call Calorimetry, or the method of measuring quantities of heat.

When heat is applied to a body it produces effects of various kinds. In most cases it raises the temperature of the body; it generally alters its volume or its pressure, and in certain cases it changes the state of the body from solid to liquid or from liquid to gaseous.

Any effect of heat may be used as a means of measuring quantities of heat by applying the principle that when two equal portions of the same substance in the same state are acted on by heat in the same way so as to produce the same effect, then the quantities of heat are equal.

We begin by choosing a standard body, and defining the standard effect of heat upon it. Thus we may choose a pound of ice at the freezing point as the standard body, and we may define as the unit of heat that quantity of heat which must be applied to this pound of ice to convert it into a pound of water still at the freezing point. This is an example of a certain change of state being used to define what is meant by a quantity of heat. This unit of heat was brought into actual use in the experiments of Lavoisier and Laplace.

In this system a quantity of heat is measured by the number of pounds (or of grammes) of ice at the freezing point which that quantity of heat would convert into water at the freezing point.

We might also employ a different system of measurement by defining a quantity of heat as measured by the number of pounds of water at the boiling point which it would convert into steam at the same temperature.

This method is frequently used in determining the amount of heat generated by the combustion of fuel.

Neither of these methods requires the use of the thermometer.

Another method, depending on the use of the thermometer, is to define as the unit of heat that quantity of heat which if applied to unit of mass (one pound or one gramme) of water at some standard temperature (that of greatest density, 39° F. or 4° C., or occasionally some temperature more convenient for laboratory work, such as 62° F. or 15° C.), will raise that water one degree (Fahrenheit or Centigrade) in temperature.

According to this method a quantity of heat is measured by the quantity of water at a standard temperature which that quantity of heat would raise one degree.

All that is assumed in these methods of measuring heat is that if it takes a certain quantity of heat to produce a certain effect on one pound of water in a certain state, then to produce the same effect on another similar pound of water will require as much heat, so that twice the quantity of heat is required for two pounds, three times for three pounds, and so on.

We have no right to assume that because a unit of heat raises a pound of water at 39° F. one degree, therefore two units of heat will raise the same pound two degrees; for the quantity of heat required to raise the water from 40° to 41° may be different from that which raised it from 39° to 40°. Indeed, it has been found by experiment that more heat is required to raise a pound of water one degree at high temperatures than at low ones.

But if we measure heat according to either of the methods already described, either by the quantity of a particular kind of matter which it can change from one easily observed state to another without altering its temperature, or by the quantity of a particular kind of matter which it can raise from one given temperature to another given temperature, we may treat quantities of heat as mathematical quantities, and add or subtract them as we please.

We have, however, in the first place to establish that the heat which by entering or leaving a body in any manner produces a given change in it is a quantity strictly comparable with that which melts a pound of ice, and differs from it only by being so many times greater or less.

In other words, we have to show that heat of all kinds, whether coming from the hand, or hot water, or steam, or red-hot iron, or a flame, or the sun, or from any othei source, can be measured in the same way, and that the quantity of each required to effect any given change, to melt a pound of ice, to boil away a pound of water, or to warm the water from one temperature to another, is the same from whatever source the heat comes.

To find whether these effects depend on anything except the quantity of heat received—for instance, if they depend in any way on the temperature of the source of heat—suppose two experiments tried. In the first a certain quantity of heat (say the heat emitted by a candle while an inch of candle is consumed) is applied directly to melt ice. In the second the same quantity of heat is applied to a piece of iron at the freezing point so as to warm it, and then the heated iron is placed in ice so as to melt a certain quantity of ice, while the iron itself is cooled to its original temperature.

If the quantity of ice melted depends on the temperature of the source from whence the heat proceeds, or on any other circumstance than the quantity of the heat, the quantity melted will differ in these two cases ; for in the first the heat comes directly from an exceedingly hot flame, and in the second the same quantity of heat comes from comparatively cool iron.

It is found by experiment that no such difference exists, and therefore heat, considered with respect to its power of warming things and changing their state, is a quantity strictly capable of measurement, and not subject to any variations in quality or in kind.

Another principle, the truth of which is established by calorimetrical experiments, is, that if a body in a given state is first heated so as to make it pass through a series of states defined by the temperature and the volume of the body in each state, and if it is then allowed to cool so as to pass in reverse order through exactly the same series of states, then the quantity of heat which entered it during the heating process is equal to that which left it during the cooling process. By those who regarded heat as a substance, and called it Caloric, this principle was regarded as self-evident, and was generally tacitly assumed. We shall show, however, that though it is true as we have stated it, yet, if the series of states during the process of heating is different from that during the process of cooling, the quantities of heat absorbed and emitted may be different. In fact heat may be generated or destroyed by certain processes, and this shows that heat is not a substance. By finding what it is produced from, and what it is reduced to, we may hope to determine the nature of heat.

In most of the cases in which we measure quantities of heat, the heat which we measure is passing out of one body into another, one of these bodies being the calorimeter itself. We assume that the quantity of heat which leaves the one body is equal to that which the other receives, provided, 1st, that neither body receives or parts with heat to any third body; and, 2ndly, that no action takes place among the bodies except the giving and receiving of heat.

The truth of this assumption may be established experimentally by taking a number of bodies at different temperatures, and determining first the quantity of heat required to be given to or taken from each separately to bring it to a certain standard temperature. If the bodies are now brought to their original temperatures, and allowed to exchange heat among themselves in any way, then the total quantity of heat required to be given to the system to bring it to the standard temperature will be found to be the same as that which would be deduced from the results in the first case.

We now proceed to describe the experimental methods by which these results may be verified, and by which quantities of heat in general may be measured.

In some of the earlier experiments of Black on the heat required to melt ice and to boil water, the heat was applied by means of a flame, and as the supply of heat was assumed to be uniform, the quantities of heat supplied were inferred to be proportional to the time during which the supply continued. A method of this kind is obviously very imperfect, and in order to make it at all accurate would need numerous precautions and auxiliary investigations with respect to the laws of the production of heat by the flame and its application to the body which is heated. Another method, also depending on the observation of time, is more worthy of confidence. We shall describe it under the name of the Method of Cooling.

ICE CALORIMETERS.

Wilcke, a Swede, was the first who employed the melting of snow to measure the heat given off by bodies in cooling. The principal difficulty in this method is to ensure that all the heat given off by the body is employed in melting the ice, and that no other heat reaches the ice so as to melt it, or escapes from the water so as to freeze it. This condition was first fulfilled by the calorimeter of Laplace and Lavoisier, of which the description is given in the Memoirs of the French Academy of Sciences for 1780. The instrument itself is preserved in the Conservatoire des Arts et Métiers at Paris.

This apparatus, which afterwards received the name of Calorimeter, consists of three vessels, one within another.

FIG. 6.

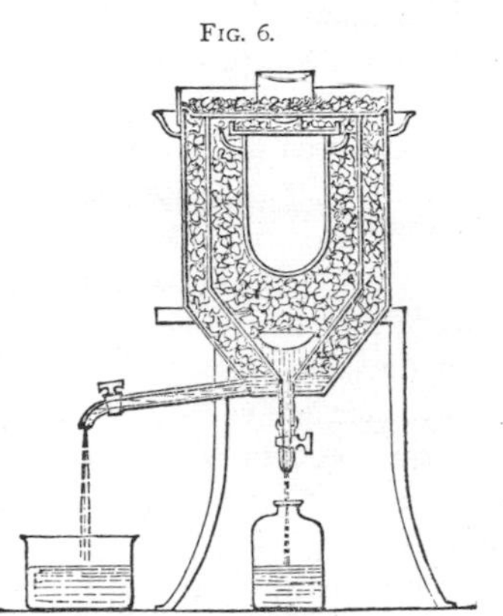

The first or innermost vessel, which we may call the receiver, is intended to hold the body from which the heat to be measured escapes. It is made of thin sheet copper, so that the heat may readily pass into the second vessel. The second vessel, or calorimeter proper, entirely surrounds the first. The lower part of the space between the two vessels is filled with broken ice at the freezing (or melting) point, and the first vessel is then covered by means of a lid, which is itself a vessel full of broken ice. When the ice melts in this vessel, whether in the lower part or in the cover of the first vessel, the water trickles down and passes through a drainer, which prevents any ice from escaping, and so runs out into a bottle set to catch it. The third vessel, which we may call the ice jacket, entirely surrounds the second, and is furnished, like the second, with an upper lid to cover the second. Both the vessel and the lid are full of broken ice at the freezing point, but the water formed by the melting of this ice is carried off to a vessel distinct from that which contains the water from the calorimeter proper.

Now, suppose that there is nothing in the receiver, and that the temperature of the surrounding air is above the freezing point. Any heat which enters the outer vessel will melt some of the ice in the jacket, and will not pass on, and no ice will be melted in the calorimeter. As long as there is ice in the jacket and in the calorimeter the temperature of both will be the same, that is, the freezing point, and therefore, by the law of equilibrium of heat, no heat will pass through the second vessel either outwards or inwards. Hence, if any ice is melted in the calorimeter, the heat which melts it must come from the receiver.

Let us next suppose the receiver at the freezing temperature; let the two lids be carefully lifted off for an instant, and a body at some higher temperature introduced into the receiver; then let the lids be quickly replaced. Heat will pass from the body through the sides of the receiver into the calorimeter, ice will be melted, and the body will be cooled, and this process will go on till the body is cooled to the freezing point, after which there will be no more ice melted.

If we measure the water produced by the melting of the ice, we may estimate the quantity of heat which escapes from the body while it cools from its original temperature to the freezing point. The receiver is at the freezing point at the beginning and at the end of the operation, so that the heating and subsequent cooling of the receiver does not influence the result.

Nothing can be more perfect than the theory and design of this apparatus. It is worthy of Laplace and of Lavoisier, and in their hands it furnished good results.

The chief inconvenience in using it arises from the fact that the water adheres to the broken ice instead of draining away from it completely, so that it is impossible to estimate accurately how much ice has really been melted.

To avoid this source of uncertainty, Sir John Herschel proposed to fill the interstices of the ice with water at the freezing point, and to estimate the quantity of ice melted by the contraction which the volume of the whole undergoes, since, as we shall afterwards see, the volume of the water is less than that of the ice from which it was formed. I am not aware that this suggestion was ever developed into an experimental method.

Bunsen,[1] independently, devised a calorimeter founded on the same principle, but in the use of which the sources of error are eliminated, and the physical constants determined with a degree of precision seldom before attained in researches of this kind.

Bunsen's calorimeter, as devised by its author, is a small instrument. The body which is to give off the heat which is to be measured is heated in a test-tube placed in a current of steam of known temperature. It is then dropped, as quickly as may be, into the test-tube T of the calorimeter, which contains water at 32° F. The body sinks to the bottom and gives off heat to the water. The heated water does not rise in the tube, for the effect of heat on water between 32° and 39° F. is to increase its density. It therefore remains surrounding the body at the bottom of the tube, and its heat can escape only by conduction either upwards through the water, or through the sides of the tube, which, being thin, afford a better channel. The tube is surrounded by ice at 32° in the calorimeter, C, so that as soon as any part of the water in the tube is raised to a higher temperature, conduction takes place through the sides, and part of the ice is melted. This will go on till everything within the tube is again reduced to 32° F., and the whole quantity of ice melted by heat *from within* is an accurate measure of the heat which the heated body gives out as it cools to 32° F.

FIG. 7.

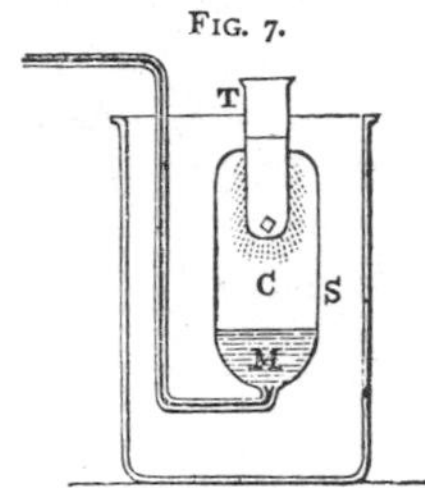

To prevent any exchange of heat between the calorimeter C and surrounding bodies, it is placed in a vessel S filled with snow gathered when new fallen and free from smoke. This

[1] *Pogg. Ann.* Sept. 1870, and *Phil. Mag.* 1871.

substance, unless the temperature of the room is below 32°, soon acquires and long maintains the temperature of 32° F.

In preparing the calorimeter, it is filled with distilled water, from which every trace of air must be expelled by a careful process of boiling. If there is air in the water, the process of freezing expels it and produces bubbles of air, the volume of which introduces an error of measurement. The lower part of the calorimeter contains mercury, and communicates with a bent tube also containing mercury. The upper part of this tube is bent horizontally, and is carefully calibrated and graduated. As the mercury and the vessel are always at the temperature 32° F., they are of constant volume, and any changes in the position of the mercury in the graduated tube are due to the melting of ice in the calorimeter, and the consequent diminution of volume of the mass of ice and water in it.

The motions of the extremity of the column of mercury being proportional to the quantities of heat emitted from the test-tube into the calorimeter, it is easy to see how quantities of heat may be compared. In fact, Bunsen has made satisfactory determinations of the specific heat of those rare metals, such as indium, of which only a few grammes have been obtained.

To prepare the calorimeter for use, ice must be formed in the calorimeter round the test-tube. For this purpose, Bunsen causes a current of alcohol, cooled below 32° by a freezing mixture, to flow to the bottom of the test-tube and up along its sides. In this way the greater part of the water in the calorimeter is soon frozen. When the apparatus has been left for a sufficient time in the vessel containing snow, the temperature of this ice rises to 32° F., and the apparatus is ready for use. A great many experiments may be made after one freezing of the water.[1]

[1] See *Pogg. Ann.* Sept. 1870, or *Phil. Mag.* 1871.

METHOD OF MIXTURE.

The second calorimetric method is usually called the Method of Mixture. This name is given to all the processes in which the quantity of heat which escapes from one body is measured by the increase of temperature it produces in another body into which it escapes. The most perfect method of ensuring that all the heat which escapes from the one body passes into the other is to mix them, but in many cases to which the method is now applied this cannot be done.

We shall illustrate this method by a few experiments, which can be performed by the student without any special apparatus. A few experiments of this kind actually performed by himself will give the student a more intelligent interest in the subject, and will give him a more lively faith in the exactness and uniformity of nature, and in the inaccuracy and uncertainty of our observations, than any reading of books, or even witnessing elaborate experiments performed by professed men of science.

I shall suppose the student to have a thermometer, the bulb of which he can immerse in the liquids of which the temperature is to be measured, and I shall suppose the graduation of the thermometer to be that of Fahrenheit, as it is the most common in this country.

To compare the effects of heat on water and on lead, take a strip of sheet lead, weighing, say, one pound, and roll it into the form of a loose spiral, so that when it is dropped into water the water may play round every part of it freely.

Take a vessel of a convenient shape, such that the roll of lead when placed in the vessel will be well covered with a pound of water.

Hang up the lead by a fine string and dip it in a saucepan of boiling water, and continue to boil it till it is thoroughly heated. While this is going on weigh out a pound of cold water in your vessel, and ascertain its temperature with the thermometer. Then lift the roll of lead out of the boiling water, hold it in the steam till the water is drained off, and immerse it as quickly as possible in the cold water in the vessel. By means of the string you may stir it about in the water so as to bring it in contact with new portions of the water, and to prevent it from giving its heat directly to the sides of the vessel.

From time to time observe the temperature of the water as indicated by the thermometer. In a few minutes the temperature of the water will cease to rise, and the experiment may then be stopped and the calculation begun.

I shall suppose (for the sake of fixing our ideas) that the temperature of the water before the hot lead was put in was 57° F., and that the final temperature, when the lead ceased to impart heat to the water, was 62° F. If we take as our unit of heat that quantity of heat which would raise a pound of water at 60° F. one degree, we have here five units of heat imparted to the water by the lead.

Since the lead was for some time in boiling water, and was afterwards held in the steam, we may assume its original temperature to be 212° (this, however, should be tested by the thermometer). During the experiment the lead cooled 150°—from 212° to 62°—and gave out, as we have seen, five units of heat to the water. Hence the difference of the heat of a pound of lead at 212° and at 62° is five units; or the same quantity of heat which will heat a pound of water five degrees from 57° to 62° will heat a pound of lead 150 degrees from 62° to 212°. If we assume, what is nearly though not exactly true, that the quantity of heat required to heat the lead is the same for each degree of rise of temperature, then we might say that to raise a pound of lead five degrees requires only one thirtieth part of the heat required to raise a pound of water five degrees.

We have thus made a comparison of the effects of heat on lead and on water. We have found that the same quantity of heat would raise a pound of lead through thirty times as many degrees as it would raise a pound of water, and we have inferred that to produce any moderate change of temperature on a pound of lead requires one-thirtieth of the heat required to produce the same change on an equal weight of water.

This comparison is expressed in scientific language by saying that the capacity of lead for heat is one-thirtieth of that of an equal weight of water.

Water is generally taken as a standard substance with which other substances are compared, and the fact which we have stated above is expressed in a still more concise manner by saying that the specific heat of lead is $\frac{1}{30}$.

The fact that when equal weights of quicksilver and water are mixed together the resulting temperature is not the mean of the temperatures of the ingredients was known to Boerhaave and Fahrenheit. Dr. Black, however, was the first to explain this phenomenon and many others by the doctrine which he established, that the effect of the same quantity of heat in raising the temperature of the body depends not only on the amount of matter in the body, but on the kind of matter of which it is formed. Dr. Irvine, Black's pupil and assistant, gave to this property of bodies the name of Capacity for Heat. The expression Specific Heat was afterwards introduced by Gadolin, of Abo, in 1784.

I think we shall secure accuracy, along with the greatest conformity to established custom, by defining these terms thus:

DEFINITION OF THE CAPACITY OF A BODY.

The capacity of a body for heat is the number of units of heat required to raise that body one degree of temperature.

We may speak of the capacity for heat of a particular thing, such as a copper vessel, in which case the capacity depends on the weight as well as on the kind of matter.

The capacity of a particular thing is often expressed by stating the quantity of water which has the same capacity.

We may also speak of the capacity for heat of a substance, such as copper, in which case we refer to unit of mass of the substance.

DEFINITION OF SPECIFIC HEAT.

The Specific Heat of a body is the ratio of the quantity of heat required to raise that body one degree to the quantity required to raise an equal weight of water one degree.

The specific heat therefore is a ratio of two quantities of the same kind, and is expressed by the same number, whatever be the units employed by the observer, and whatever thermometric scale he adopts.

It is very important to bear in mind that these phrases mean neither more nor less than what is stated in these definitions.

Irvine, who contributed greatly to establish the fact that the quantity of heat which enters or leaves a body depends on its capacity for heat multiplied by the number of degrees through which its temperature rises or falls, went on to assume that the whole quantity of heat in a body is equal to its capacity multiplied by the total temperature of the body, reckoned from a point which he called the absolute zero. This is equivalent to the assumption that the capacity of the body remains the same from the given temperature downwards to this absolute zero. The truth of such an assumption could never be proved by experiment, and its falsehood is easily established by showing that the specific heat of most liquid and solid substances is different at different temperatures.

The results which Irvine, and others long after him, deduced by calculations founded on this assumption are not only of no value, but are shown to be so by their inconsistency with each other.

We shall now return to the consideration of the experiment

with the lead and water, in order to show how it can be made more accurate by attending to all the circumstances of the case. I have purposely avoided doing so at first, as my object was to illustrate the meaning of 'Specific Heat.'

In the former description of the experiment it was assumed, not only that all the heat which escapes from the lead enters the water in the vessel, but that it remains in the water till the conclusion of the experiment, when the temperatures of the lead and water have become equalised.

The latter part of this assumption cannot be quite true, for the water must be contained in a vessel of some kind, and must communicate some of its heat to this vessel, and also must lose heat at its upper surface by evaporation, &c.

If we could form the vessel of a perfect non-conductor of heat, this loss of heat from the water would not occur; but no substance of which a vessel can be formed can be considered even approximately a non-conductor of heat; and if we use a vessel which is merely a slow conductor of heat, it is very difficult, even by the most elaborate calculations, to determine how much heat is taken up by the vessel itself during the experiment.

A better plan is to use a vessel which is a very good conductor of heat, but of which the capacity for heat is small, such as a thin copper or silver vessel, and to prevent this vessel from parting rapidly with its heat by polishing its outer surface, and not allowing it to touch any large mass of metal, but rather giving it slender supports and placing it within a metal vessel having its inner surface polished.

In this way we shall ensure that the heat shall be quickly distributed between the water and the vessel, and may consider their temperatures at all times nearly equal, while the loss of heat from the vessel will take place slowly and at a rate which may be calculated when we know the temperature of the vessel and of the air outside.

For this purpose, if we intended to make a very elaborate

experiment, we should in the first place determine the capacity for heat of the vessel by a separate experiment, and then we should put into the vessel about a pound of warm water and determine its temperature from minute to minute, while at the same time we observe with another thermometer the temperature of the air in the room. In this way we should obtain a set of observations from which we might deduce the *rate of cooling* for different temperatures, and compute the rate of cooling when the vessel is one, two, three, &c., degrees hotter than the air; and then, knowing the temperature of the vessel at various stages of the experiment for finding the specific heat of lead, we should be able to calculate the loss of heat from the vessel due to the cooling during the continuance of the experiment.

But a much simpler method of getting rid of these difficulties is by the method of making two experiments—the first with the lead which we have described, and the second with hot water, in which we endeavour to make the circumstances which cause the loss of heat as similar as we can to those in the case of the lead.

For instance, if we suppose that the specific gravity of lead is about eleven times that of water, if instead of a pound of lead we use one-eleventh of a pound of water, the bulk of the water will be the same as that of the lead, and the depth of the water in the vessel will be equally increased by the lead and the water.

If we also suppose that the specific heat of lead is one-thirtieth of that of water, then the heat given out by a pound of lead in cooling 150° will be equal to the heat given out by one-eleventh of a pound of water in cooling 55°.

Hence, if we take one-eleventh of a pound of water at 55° above 62°, that is at 117°, and pour it into the vessel with the water as before at 57°, we may expect that the level of the water will rise as much as when the hot lead was put in, and that the temperature will also rise to about the same degree. The only difference between the experiments, as far as the loss of heat is concerned, is, that the warm water will raise the temperature of the cold water in a much shorter time than the hot lead did, so that if we observe the temperature at the same time after the mixture in both cases, the loss by cooling will be greater with the warm water than with the hot lead.

In this way we may get rid of the chief part of the difficulty of many experiments of comparison. Instead of making one experiment, in which the cooling of the lead is compared with the heating of the water and the vessel, including an unknown loss of heat from the outside of the vessel, we make two experiments, in which the heating of the vessel and the total loss of heat shall be as nearly as possible the same, but in which the heat is furnished in the one case by hot lead, and in the other by warm water. The student may compare this method with the method of double weighing invented by Père Amiot, but commonly known as Borda's method, in which first the body to be weighed, and then the weights, are placed in the same scale, and weighed against the same counterpoise.

We shall illustrate this method by finding the effect of steam in heating water, and comparing it with that of hot water. Take a kettle, and make the lid tight with a little flour and water, and adapt a short india-rubber tube to the spout, and a tin or glass nozzle to the tube. Make the water in the kettle boil, and when the steam comes freely through the nozzle dip it in cold water, and you will satisfy yourself that the steam is rapidly condensed, every bubble of steam as it issues collapsing with a sharp rattling noise.

Having made yourself familiar with the general nature of the experiment of the condensation of steam, you may proceed to measure the heat given out to the water. For this purpose, put some cold water in your vessel, say about three-quarters of a pound. Weigh the vessel and water carefully, and observe the temperature of the water; then, while the steam flows freely from the nozzle, condense steam in the water for a short time, and remove the nozzle ; observe the temperature and weigh the water in its vessel again, taking note of the time of the experiment.

Let us suppose the original weight . .	5,000 grains
Weight after the condensation of steam .	5,100 grains
Hence the weight of steam condensed is .	100 grains
Temperature of water at first . . .	55° F.
Temperature at the end of experiment .	77° F.
Rise of temperature	22°

Let us now make a second experiment, as like the first as we can, only differing from it by the use of hot water instead of steam to produce the rise of temperature.

It is impossible in practice to ensure that everything shall be exactly the same, but after a few trials we may select a method which will nearly, if not quite, fulfil the conditions.

Thus it is easy to bring the vessel and cold water to the same weight as before, namely, 5,000 grains ; but we shall suppose the temperature now to be 56° F. instead of 55°. We now pour in water at 176° F. gradually, so as to make this experiment last about as long as the first, and we find that the temperature is now 76°, and the weight 6,000 grains. Hence 1,000 grains of water cooling 100° raise the vessel and its contents 22°.

Assuming that the specific heat of water is the same at all temperatures, which is nearly, though by no means exactly, true, the quantity of heat given out by the water in the second experiment is equal to what would raise 100,000 grains of water one degree.

In the experiment with the steam the temperatures were nearly though not exactly equal, but the rise of temperature was greater in the proportion of 22 to 20. Hence we may conclude that the quantity of heat which produced this heating effect in the experiment with steam was greater than in the experiment with water in the same proportion. This makes the heat given out by the steam equal to that which would raise 110,000 grains of water one degree.

This was done by the condensation and subsequent cooling of 100 grains of steam. Let us begin with the heat given out by the 100 grains of water at 212° F., into which the steam is condensed. It is cooled from 212° to 77° or 135°, and gives out therefore an amount of heat which would raise 13,500 grains of water one degree. But the whole effect was 110,000, so that there is an amount of heat which would raise 96,500 grains of water one degree, which must be given out during the condensation of the steam, and before the cooling begins. Hence each grain of steam in condensing gives out as much heat as would raise 965 grains of water 1° F. or 536 grains 1° Centigrade.

The fact that steam at the boiling point gives out a large quantity of heat when it is condensed into water which is still at the same temperature, and the converse fact that in order to convert water at the boiling temperature into steam of the same temperature a large quantity of heat must be communicated to it, was first clearly established by Black in 1757.

He expressed it by saying that the latent heat of steam is 965° F., and this form of expression is still in use, and we should take it to mean neither more nor less than what we have just stated.

Black, however, and many of his followers, supposed heat to be a substance which when it makes a thing hot is sensible, but which when it is not perceived by the hand or the thermometer still exists in the body in a latent or concealed state. Black supposed that the difference between boiling water and steam is, that steam contains a great deal more caloric than the hot water, so that it may be considered a compound of water and caloric; but, since this additional caloric produces no effect on the temperature, but lurks concealed in the steam ready to appear when it is condensed, he called this part of the heat latent heat.

In considering the scientific value of Black's discovery of

latent heat, and of his mode of expressing it, we should recollect that Black himself in 1754 was the discoverer of the fact that the bubbles formed when marble is put into an acid consist of a real substance different from air, which, when free, is similar to air in appearance, but when fixed may exist in liquids and in solids. This substance, which we now call carbonic acid, Black called fixed air, and this was the first gaseous body distinctly recognised as such. Other airs or gases were afterwards discovered, and the impulse given to chemistry was so great, on account of the extension of the science to these attenuated bodies, that most philosophers of the time were of opinion that heat, light, electricity, and magnetism, if not the vital force itself, would sooner or later be added to the list. Observing, however, that the gases could be weighed, while the presence of these other agents could not be detected by the balance, those who admitted them to the rank of substances called them imponderable substances, and sometimes, on account of their mobility, imponderable fluids.

The analogy between the free and fixed states of carbonic acid and the sensible and latent states of heat encouraged the growth of materialistic phrases as applied to heat; and it is evident that the same way of thinking led electricians to the notion of disguised or dissimulated electricity, a notion which survives even yet, and which is not so easily stripped of its erroneous connotation as the phrase 'latent heat.'

It is worthy of remark that Cavendish, though one of the greatest chemical discoverers of his time, would not accept the phrase 'latent heat.' He prefers to speak of the generation of heat when steam is condensed, a phrase inconsistent with the notion that heat is matter, and objects to Black's term as relating 'to an hypothesis depending on the supposition that the heat of bodies is owing to their containing more or less of a substance called the matter of heat; and, as I think Sir Isaac Newton's opinion that heat consists in the internal motion of the particles of bodies much the most probable, I chose to use the expression, "heat is generated."'[1]

We shall not now be in danger of any error if we use latent heat as an expression meaning neither more nor less than this :

DEFINITION.—*Latent heat is the quantity of heat which must be communicated to a body in a given state in order to convert it into another state without changing its temperature.*

We here recognise the fact that heat when applied to a body may act in two ways—by changing its state, or by raising its temperature—and that in certain cases it may act by changing the state without increasing the temperature.

The most important cases in which heat is thus employed are—

1. The conversion of solids into liquids. This is called melting or fusion. In the reverse process of freezing or solidification heat must be allowed to escape from the body to an equal amount.

2. The conversion of liquids (or solids) into the gaseous state. This is called evaporation, and its reverse condensation.

3. When a gas expands, in order to maintain the temperature constant, heat must be communicated to it, and this, when properly defined, may be called the latent heat of expansion.

4. There are many chemical changes during which heat is generated or disappears.

In all these cases the quantity of heat which enters or leaves the body may be measured, and in order to express the result of this measurement in a convenient form, we may call it the latent heat required for a given change in the substance.

We must carefully remember that all that we know about heat is what occurs when it passes from one body to another,

[1] *Phil. Trans.* 1783, quoted by Forbes. Dissertation VI. *Encyc. Brit.*

and that we must not assume that after heat has entered a substance it exists in the form of heat within that substance. That we have no right to make such an assumption will be abundantly shown by the demonstration that heat may be transformed into and may be produced from something which is not heat.

Regnault's method of passing large quantities of the substance through the calorimeter will be described in treating of the properties of gases, and the Method of Cooling will be considered in the chapter on Radiation.

CHAPTER IV.

ELEMENTARY DYNAMICAL PRINCIPLES.

IN the first part of this treatise we have confined ourselves to the explanation of the method of ascertaining the temperature of bodies, which we call thermometry, and the method of measuring the quantity of heat which enters or leaves a body, and this we call calorimetry. Both of these are required in order to study the effects of heat upon bodies; but we cannot complete this study without making measurements of a mechanical kind, because heat and mechanical force may act on the same body, and the actual result depends on both actions. I propose, therefore, to recall to the student's memory some of those dynamical principles which he ought to bring with him to the study of heat, and which are necessary when he passes from purely thermal phenomena, such as we have considered, to phenomena involving pressure, expansion, &c., and which will enable him afterwards to proceed to the study of thermodynamics proper, in which the relations of thermal phenomena among themselves are deduced from purely dynamical principles.

The most important step in the progress of every

science is the measurement of quantities. Those whose curiosity is satisfied with observing what happens have occasionally done service by directing the attention of others to the phenomena they have seen ; but it is to those who endeavour to find out how much there is of anything that we owe all the great advances in our knowledge.

Thus every science has some instrument of precision, which may be taken as a material type of that science which it has advanced, by enabling observers to express their results as measured quantities. In astronomy we have the divided circle, in chemistry the balance, in heat the thermometer, while the whole system of civilised life may be fitly symbolised by a foot rule, a set of weights, and a clock. I shall, therefore, make a few remarks on the measurement of quantities.

Every quantity is expressed by a phrase consisting of two components, one of these being the name of a number, and the other the name of a thing of the same kind as the quantity to be expressed, but of a certain magnitude agreed on among men as a standard or unit.

Thus we speak of two days, of forty-eight hours.

Each of these expressions has a numerical part and a denominational part, the numerical part being a number, whole or fractional, and the denominational part being the name of the thing, which is to be taken as many times as is indicated by the number.

If the numerical part is the number *one*, then the quantity is the standard quantity itself, as when we say one pound, or one inch, or one day. A quantity of which the numerical part is unity is called a unit. When the numerical part is some other number, the quantity is still said to be referred to, or to be expressed in terms of that quantity which would be denoted if the number were *one*, and which is called the unit.

In all cases the unit is a quantity of the same kind as the quantity which is expressed by means of it.

In many cases several units of the same kind are in use,

as miles, yards, feet, and inches, as measures of length ; cubic yards, gallons, and fluid ounces, as measures of capacity; besides the endless variety of units which have been adopted by different nations, and by different districts and different trades in the same nation.

When a quantity given in terms of one unit has to be expressed in terms of another, we find the number of times the second unit is contained in the first, and multiply this by the given number.

Hence the numerical part of the expression of the same quantity varies inversely as the unit in which it is to be expressed, as in the example, two days and forty-eight hours, which mean the same thing.

There are many quantities which can be defined in terms of standard quantities of a different kind. In this case we make use of derived units. For instance, as soon as we have fixed on a measure of length, we may define by means of it not only all lengths, but also the area of any surface, and the content of any space. For this purpose, if the foot is the unit of length, we construct, by Euclid I. 46, a square whose side is a foot, and express all areas in terms of this square foot, and by constructing a cube whose edge is a foot we have defined a cubic foot as a unit of capacity.

We also express velocities in miles an hour, or feet in a second, &c.

In fact, all quantities with which we have to do in dynamics may be expressed in terms of units derived by definition from the three fundamental units—of Length, Mass, and Time.

STANDARD OF LENGTH.

It is so important to mankind that these units should be well defined that in all civilised nations they are defined by the State with reference to material standards, which are preserved with the utmost care. For instance, in this country it was enacted by Parliament [1] 'that the straight line or

[1] 18 & 19 Vict. c. 72, July 30, 1855.

distance between the centres of the transverse lines in the two gold plugs in the bronze bar deposited in the office of the Exchequer shall be the genuine standard yard at 62° F., and if lost it shall be replaced by means of its copies.'

The authorised copies here referred to are those which are preserved at the Royal Mint, the Royal Society of London, the Royal Observatory at Greenwich, and the New Palace at Westminster. Other copies have been made with great care, and with these all measures of length must be compared.

The length of the Parliamentary standard was chosen so as to be as nearly as possible equal to that of the best standard yards formerly used in England. The State, therefore, endeavoured to maintain the standard of its ancient magnitude, and by its authority it has defined the actual magnitude of this standard with all the precision of which modern science is capable.

The mètre derives its authority as a standard from a law of the French Republic in 1795.

It is defined to be the distance between the ends of a rod of platinum made by Borda,[1] the rod being at the temperature of melting ice. This distance was chosen without reference to any former measures used in France. It was intended to be a universal and not a national measure, and was derived from Delambre and Mechain's measurement of the size of the earth. The distance measured along the earth's surface from the pole to the equator is nearly ten million of mètres. If, however, in the progress of geodesy, a different result should be obtained from that of Delambre, the mètre will not be altered, but the new result will be expressed in the old mètres. The authorised standard of length is therefore not the terrestrial globe, but Borda's

[1] Mètre conforme à la loi du 18 Germinal, an III. Présenté le 4 Messidor, an VII.

platinum rod, which is much more likely to be accurately measured.

The value of the French system of measures does not depend so much on the absolute values of the units adopted as on the fact that all the units of the same kind are connected together by a decimal system of multiplication and division, so that the whole system, under the name of the metrical system, is rapidly gaining ground even in countries where the old national system has been carefully defined.

The mètre is 39·37043 British inches.

STANDARD OF MASS.

By the Act above cited a weight of platinum marked 'P. S, 1844, 1 lb.,' deposited in the office of the Exchequer, 'shall be the legal and genuine standard measure of weight, and shall be and be denominated the Imperial Standard Pound Avoirdupois, and shall be deemed to be the only standard measure of weight from which all other weights and other measures having reference to weight shall be derived, computed, and ascertained, and one equal seven-thousandth part of such pound avoirdupois shall be a grain, and five thousand seven hundred and sixty such grains shall be and be deemed to be a pound troy. If at any time hereafter the said Imperial Standard Pound Avoirdupois be lost or in any manner destroyed, defaced, or otherwise injured, the Commissioners of Her Majesty's Treasury may cause the same to be restored by reference to or adoption of any of the copies aforesaid,[1] or such of them as may remain available for that purpose.'

The construction of this standard was entrusted to Professor W. H. Miller, who has given an account of the methods employed in a paper,[2] which may be here referred to as a model of scientific accuracy.

[1] In the same places as the Standards of Length.

[2] *Phil. Trans.* 1856, p. 753.

The French standard of mass is the Kilogramme des Archives, made of platinum by Borda, and is intended to represent the mass of a cubic décimètre of distilled water at the temperature 4° C.

The actual determination of the density of water is an operation which requires great care, and the differences between the results obtained by the most skilful observers, though small, are a thousand times greater than the differences of the results of a comparison of standards by weighing them. The differences of the values of the density of water as found by careful observers are as much as a thousandth part of the whole, whereas the method of weighing admits of an accuracy of within one part in five millions.

Hence the French standards, though originally formed to represent certain natural quantities, must be now considered as arbitrary standards, of which copies are to be taken by direct comparison. The French or metric system has the advantage of a uniform application of the decimal method, and it is also in many cases convenient to remember that a cubic mètre of water is a tonne, a cubic décimètre a kilogramme, a cubic centimètre a gramme, and a cubic millimètre a milligramme, the water being at its maximum density or at about 4° C.

In 1826 the British standard of mass was defined by saying that a cubic inch of water at 62° F. contains 252·458 grains, and though this is no longer a legal definition, we may take it as a rough statement of a fact, that a cubic inch of water weighs *about* 252·5 grains, a cubic foot *about* 1,000 ounces avoirdupois, and a cubic yard *about* three-quarters of a ton. Of these estimates the second is the furthest from the truth.

Professor Miller has compared the British and French standards, and finds the Kilogramme des Archives equal to 15432·34874 grains.

From these legal definitions it will be seen that what is generally called a standard of weight is a certain piece of platinum—that is, a particular body the quantity of matter in which is taken and defined by the State to be a pound or a kilogramme.

The weight strictly so called—that is, the tendency of this body to move downwards—is not invariable, for it depends on the part of the world where it is placed, its weight being greater at the poles than at the equator, and greater at the level of the sea than at the top of a mountain.

What is really invariable is the quantity of matter in the body, or what is called in scientific language the mass of the body, and even in commercial transactions what is generally aimed at in weighing goods is to estimate the quantity of matter, and not to determine the force with which they tend downwards.

In fact, the only occasions in common life in which it is required to estimate weight considered as a force is when we have to determine the strength required to lift or carry things, or when we have to make a structure strong enough to support their weight. In all other cases the word weight must be understood to mean *the quantity of the thing as determined by the process of weighing against 'standard weights.'*

As a great deal of confusion prevails on this subject in ordinary language, and still greater confusion has been introduced into books on mechanics by the notion that a pound is a certain force, instead of being, as we have seen, a certain piece of platinum, or a piece of any other kind of matter equal in mass to the piece of platinum, I have thought it worth while to spend some time in defining accurately what is meant by a pound and a kilogramme.

ON THE UNIT OF TIME.

All nations derive their measures of time from the apparent motions of the heavenly bodies. The motion of rotation of the earth about its axis is very nearly indeed uniform, and the measure of time in which one day is equal to the time of revolution of the earth about its axis, or more exactly to the interval between successive transits of the first point of Aries, is used by astronomers under the name of sidereal time.

Solar time is that which is given by a sun-dial, and is not uniform. A uniform measure of time, agreeing with solar time in the long run, is called mean solar time, and is that which is given by a correct clock. A solar day is longer than a sidereal day. In all physical researches mean solar time is employed, and one second is generally taken as the unit of time.

The evidence upon which we form the conclusion that two different portions of time are or are not equal can only be appreciated by those who have mastered the principles of dynamical reasoning. I can only here assert that the comparison, for example, of the length of a day at present with the length of a day 3,000 years ago is by no means an unfruitful enquiry, and that the relative length of these days may be determined to within a small fraction of a second. This shows that time, though we conceive it merely as the succession of our states of consciousness, is capable of measurement, independently, not only of our mental states, but of any particular phenomenon whatever.

ON MEASUREMENTS FOUNDED ON THE THREE FUNDAMENTAL UNITS.

In the measurement of quantities differing in kind from the three units, we may either adopt a new unit independently for each new quantity, or we may endeavour to define a unit of the proper kind from the fundamental units. In the latter case we are said to use a system of units. For instance, if we have adopted the foot as a unit of length, the systematic unit of capacity is the cubic foot.

The gallon, which is a legal measure in this country, is unsystematic considered as a measure of capacity, as it contains the awkward number of 277·274 cubic inches. The gallon, however, is never tested by a direct measurement of its cubic contents, but by the condition that it must contain ten pounds of water at 62° F.

DEFINITION OF DENSITY.—*The density of a body is measured by the number of units of mass in unit of volume of the substance.*

For instance, if the foot and the pound be taken as fundamental units, then the density of anything is the number of pounds in a cubic foot. The density of water is about 62·5 pounds to the cubic foot. In the metric system, the density of water is one tonne to the stère, one kilogramme to the litre, one gramme to the cubic centimètre, and one milligramme to the cubic millimètre.

We shall sometimes have to use the word rarity, to signify the inverse of density, that is, the volume of unit of mass of a substance.

DEFINITION OF SPECIFIC GRAVITY.—*The specific gravity of a body is the ratio of its density to that of some standard substance, generally water.*

Since the specific gravity of a body is the ratio of two things of the same kind, it is a numerical quantity, and has the same value, whatever national units are employed by those who determine it. Thus, if we say that the specific gravity of mercury is about 13·5, we state that mercury is about thirteen and a half times heavier than an equal bulk of water, and this fact is independent of the way in which we measure either the mass or the volume of the liquids.

DEFINITION OF UNIFORM VELOCITY.—*The velocity of a body moving uniformly is measured by the number of units of length travelled over in unit of time.*

Thus we speak of a velocity of so many feet or mètres per second.

DEFINITION OF MOMENTUM.—*The momentum of a body is measured by the product of the velocity of the body into the number of units of mass in the body*

DEFINITION OF FORCE.—*Force is whatever changes or tends to change the motion of a body by altering either its direction or its magnitude; and a force acting on a body is measured by the momentum it produces in its own direction in unit of time.*

For instance, the weight of a pound acting on the pound itself at London would, if it acted for a second, produce a velocity of about 32·1889 feet per second in the mass itself, which is one pound. Hence, in the British system the weight of one pound at London is 32·1889 British units of force.

In Paris the weight of a gramme allowed to fall freely for one second would generate in the gramme a velocity of 9·80868 mètres per second, so that the weight of a gramme at Paris is 9·80868 metrical units of force.

It is so convenient, especially when all our experiments are conducted in the same place, to express forces in terms of the weight of a pound or a gramme, that in all countries the first measurements of forces were made in this way, and a force was described as a force of so many pounds' weight or grammes' weight. It was only after the measurements of forces made by persons in different parts of the world had to be compared that it was found that the weight of a pound or a gramme is different in different places, and depends on the intensity of gravitation, or the attraction of the earth; so that for purposes of accurate comparison all forces must be reduced to absolute or dynamical measure as explained above. We shall distinguish the measure by comparison with weight as the *gravitation* measure of force. To reduce forces expressed in gravitation measure to absolute measure, we must multiply the number denoting the force in gravitation measure by the value of the force of gravity expressed in the same metrical system. The value of the force of gravity is a very important number in all scientific calculations, and it is generally denoted by the letter g. The quantity g may be defined in any of the following ways, which are all equivalent :

g is a number expressing the velocity produced in a falling body in unit of time.

g is a number expressing twice the distance through which a body falls in unit of time.

g is a number expressing the weight of unit of mass in absolute measure.

The value of g is generally determined at any place by experiments with the pendulum. These experiments require great care, and the description of them does not belong to our present subject. The value of g may be found with sufficient accuracy for the present state of science by means of the formula,

$$g = G\,(1 - 0{\cdot}0025659 \cos 2\lambda)\left\{1 - \left(2 - \frac{3}{2}\frac{\rho'}{\rho}\right)\frac{z}{r}\right\}$$

In this formula, G is the force of gravity at the mean level of the sea in latitude 45°:

$$G = 32{\cdot}1703 \text{ feet, or } 9{\cdot}80533 \text{ mètres.}$$

λ is the latitude of the place. The formula shows that the force of gravity at the level of the sea increases from the equator to the poles. The last factor of the formula is intended to represent the effect of the height of the place of observation above the level of the sea in diminishing the force of gravity. If the observations were carried on in a balloon, or on the top of a tower, the force of gravity would vary, according to Newton's law, inversely as the square of the distance from the centre of the earth, but since observations are usually made on or near the surface of a part of the earth which is above the level of the sea, the effect of the attraction of the ground which is raised above the sea-level must be taken into account. This is done by a method investigated by Poisson,[1] and expressed in the last factor of the above formula.

The symbol ρ represents the mean density of the whole earth, which is probably about $5\frac{1}{2}$ times that of water. ρ'

[1] *Traité de Mécanique*, t. ii. p. 629.

represents the mean density of the ground just below the place of observation, which may be taken at about $2\frac{1}{2}$ times the density of water, so that we may write

$$2 - \frac{3}{2}\frac{\rho'}{\rho} = 1 \cdot 32 \text{ nearly.}$$

z is the height of the place above the level of the sea, in feet or mètres, and r is the radius of the earth :

$$r = 20{,}886{,}852 \text{ feet, or } 6{,}366{,}198 \text{ mètres.}$$

For rough purposes it is sufficient to remember that in Britain the force of gravity is about 32·2 feet, and in France about 9·8 mètres.

The reason why, in all accurate measurements, we have to take account of the variation of the force of gravity in different places is, that the absolute value of any force, such as the pressure of air of a given density and temperature, depends entirely on the properties of air, and not on the force of gravity at the place of observation. If, therefore, this pressure has been observed in gravitation measure, that is, in pounds on the square inch, or in inches of mercury, or in any way in which the weight of some substance is made to furnish the measure of the pressure, then the results so obtained will be true only as long as the force of gravity is the same, and will not be true without correction at a place in a different latitude from the place of observation. Hence the use of reducing all measures of force to absolute measure.

In a rude age, before the invention of means for overcoming friction, the weight of bodies formed the chief obstacle to setting them in motion. It was only after some progress had been made in the art of throwing missiles, and in the use of wheel-carriages and floating vessels, that men's minds became practically impressed with the idea of mass as distinguished from weight. Accordingly, while almost all the metaphysicians who discussed the qualities of matter assigned a prominent place to weight among the primary qualities, few or none of them perceived that the sole unalterable property of matter is its *mass*. At the revival of science this property was expressed by the phrase 'the inertia of matter;' but while the men of science understood by this term the tendency of the body to persevere in its state of motion (or rest), and considered it a measurable quantity, those philosophers who were unacquainted with science understood inertia in its literal sense as a quality—mere want of activity or laziness.

Even to this day those who are not practically familiar with the free motion of large masses, though they all admit the truth of dynamical principles, yet feel little repugnance in accepting the theory known as Boscovich's—that substances are composed of a system of points, which are mere centres of force, attracting or repelling each other. It is probable that many qualities of bodies might be explained on this supposition, but no arrangement of centres of force, however complicated, could account for the fact that a body requires a certain force to produce in it a certain change of motion, which fact we express by saying that the body has a certain measurable mass. No part of this mass can be due to the existence of the supposed centres of force.

I therefore recommend to the student that he should impress his mind with the idea of mass by a few experiments, such as setting in motion a grindstone or a well-balanced wheel, and then endeavouring to stop it, twirling a long pole, &c., till he comes to associate a set of acts and sensations with the scientific doctrines of dynamics, and he will never afterwards be in any danger of loose ideas on these subjects. He should also read Faraday's essay on Mental Inertia,[1] which will impress him with the proper metaphorical use of the phrase to express, not laziness, but habitude.

[1] *Life*, by Dr. Bence Jones, vol. i. p. 268.

ON WORK AND ENERGY.

Work is done when resistance is overcome, and the quantity of work done is measured by the product of the resisting force and the distance through which that force is overcome.

Thus, if a body whose mass is one pound is lifted one foot high in opposition to the force of gravity, a certain amount of work is done, and this quantity is known among engineers as a foot-pound.

If a body whose mass is twenty pounds is lifted ten feet, this might be done by taking one of the pounds and raising it first one foot and then another till it had risen ten feet, and then doing the same with each of the remaining pounds, so that the quantity of work called a foot-pound is performed 200 times in raising twenty pounds ten feet. Hence the work done in lifting a body is found by multiplying the weight of the body in pounds by the height in feet. The result is the work in foot-pounds.

The foot-pound is a *gravitation* measure, depending on the force of gravity at the place. To reduce it to absolute measure we must multiply the number of foot-pounds by the force of gravity at the place.

The work done when we raise a heavy body is done in overcoming the attraction of the earth. Work is also done when we draw asunder two magnets which attract each other, when we draw out an elastic cord, when we compress air, and, in general, when we apply force to anything which moves in the direction of the force.

There is one case of the application of force to a moving body which is of great importance, namely, when the force is employed in changing the velocity of the body.

Suppose a body whose mass is M (M pounds or M grammes) to be moving in a certain direction with a velocity which we shall call v, and let a force, which we shall call F, be applied to the body in the direction of its motion. Let us consider the effect of this force acting on the body for a very small time T, during which the body moves through the space s, and at the end of which its velocity is v'.

To ascertain the magnitude of the force F, let us consider the momentum which it produces in the body, and the time during which the momentum is produced.

The momentum of the beginning of the time T was Mv, and at the end of the time T it was Mv', so that the momentum produced by the force F acting for the time T is $Mv' - Mv$.

But since forces are measured by the momentum produced in unit of time, the momentum produced by F in one unit of time is F, and the momentum produced by F in T units of time is FT. Since the two values are equal,

$$FT = M(v' - v).$$

This is one form of the fundamental equation of dynamics. If we define the impulse of a force as the average value of the force multiplied by the time during which it acts, then this equation may be expressed in words by saying that the impulse of a force is equal to the momentum produced by it.

We have next to find s, the space described by the body during the time T. If the velocity had been uniform, the space described would have been the product of the time by the velocity. When the velocity is not uniform the time must be multiplied by the mean or average velocity to get the space described. In both these cases in which average force or average velocity is mentioned, the time is supposed to be subdivided into a number of equal parts, and the average is taken of the force or of the velocity for all these divisions of the time. In the present case, in which the time considered is so small that the change of velocity is also small, the average velocity during the time T may be taken as the arithmetical mean of the velocities at the beginning and at the end of the time, or $\frac{1}{2}(v + v')$.

Hence the space described is

$$s = \tfrac{1}{2}(v + v')T.$$

This may be considered as a kinematical equation, since it depends on the nature of motion only, and not on that of the moving body.

If we multiply together these two equations we get

$$FTs = \tfrac{1}{2}M(v'^2 - v^2)T;$$

and if we divide by T we find

$$Fs = \tfrac{1}{2}Mv'^2 - \tfrac{1}{2}Mv^2.$$

Now Fs is the work done by the force F acting on the body while it moves in the direction of F through a space s. If we also denote $\tfrac{1}{2}Mv^2$, the mass of the body multiplied by half the square of its velocity, by the expression *the kinetic energy of the body*, then $\tfrac{1}{2}Mv'^2$ will be the kinetic energy after the action of the force F through a space s.

We may now express the equation in words by saying that the work done by the force F in setting the body in motion is measured by the increase of kinetic energy during the time that the force acts.

We have proved that this is true when the interval of time during which the force acts is so small that we may consider the mean velocity during that time as equal to the arithmetical mean of the velocities at the beginning and end of the time. This assumption, which is exactly true when the force is uniform, is approximately true in every case when the time considered is small enough.

By dividing the whole time of action of the force into small parts, and proving that in each of these the work done by the force is equal to the increase of kinetic energy of the body, we may, by adding the different portions of the work and the different increments of energy, arrive at the result that the total work done by the force is equal to the total increase of kinetic energy.

If the force acts on the body in the direction opposite to the motion, the kinetic energy of the body will be diminished instead of increased, and the force, instead of doing work on the body, will be a resistance which the body in its motion overcomes. Hence a moving body can do work in overcoming resistance as long as it is in motion, and the work done by the moving body is equal to the diminution of its kinetic energy, till, when the body is brought to rest, the whole work it has done is equal to the whole kinetic energy which it had at first.

We now see the appropriateness of the name kinetic energy, which we have hitherto used merely as a name for the product $\frac{1}{2}Mv^2$. For the energy of a body may be defined as the capacity which it has of doing work, and is measured by the quantity of work which it can do. The kinetic energy of a body is the energy which it has in virtue of being in *motion*, and we have just shown that its value may be found by multiplying the mass of the body by half the square of the velocity.

In our investigation we have, for the sake of simplicity, supposed the force to act in the same direction as the motion. To make the proof perfectly general, as it is given in treatises on dynamics, we have only to resolve the actual force into two parts, one in the direction of the motion and the other at right angles to it, and to observe that the part at right angles to the motion can neither do any work on the body nor change the velocity or the kinetic energy, so that the whole effect, whether of work or of alteration of kinetic energy, depends on the part of the force which is in the direction of the motion.

The student, if not familiar with this subject, should refer to some treatise on dynamics, and compare the investigation there given with the outline of the reasoning given above. Our object at present is to fix in our minds what is meant by Work and Energy.

The great importance of giving a name to the quantity which we call Kinetic Energy seems to have been first recognised by Leibnitz, who gave to the product of the mass by the square of the velocity the name of Vis Viva. This is twice the kinetic energy.

Newton, in a scholium to his Third Law of Motion, has stated the relation between work and kinetic energy in a manner so perfect that it cannot be improved, but at the same time with so little apparent effort or desire to attract attention that no one seems to have been struck with the great importance of the passage till it was pointed out recently by Thomson and Tait.

The use of the term Energy, in a scientific sense, to express the quantity of work a body can do, was introduced by Dr. Young (' Lectures on Natural Philosophy,' Lecture VIII.).

The energy of a system of bodies acting on one another with forces depending on their relative positions is due partly to their motion, and partly to their relative position.

That part which is due to their motion was called Actual Energy by Rankine, and Kinetic Energy by Thomson and Tait.

That part which is due to their relative position depends upon the work which the various forces would do if the bodies were to yield to the action of these forces. This is called the Sum of the Tensions by Helmholtz, in his celebrated memoir on the 'Conservation of Force.'[1] Thomson called it Statical Energy, and Rankine introduced the term Potential Energy, a very felicitous name, since it not only signifies the energy which the system has not in possession, but only has the power to acquire, but it also indicates that it is to be found from what is called (on other grounds) the Potential Function.

Thus when a heavy body has been lifted to a certain height above the earth's surface, the system of two bodies, it and the earth, have potential energy equal to the work which would be done if the heavy body were allowed to descend till it is stopped by the surface of the earth.

If the body were allowed to fall freely, it would acquire

[1] Berlin, 1847. Translated in Taylor's *Scientific Memoirs*, Feb. 1853.

velocity, and the kinetic energy acquired would be exactly equal to the potential energy lost in the same time.

It is proved in treatises on dynamics that if, in any system of bodies, the force which acts between any two bodies is in the line joining them, and depends only on their distance, and not on the way in which they are moving at the time, then if no other forces act on the system, the sum of the potential and kinetic energy of all the bodies of the system will always remain the same.

This principle is called the Principle of the Conservation of Energy; it is of great importance in all branches of science, and the recent advances in the science of heat have been chiefly due to the application of this principle.

We cannot indeed assume, without evidence of a satisfactory nature, that the mutual action between any two parts of a real body must always be in the line joining them, and must depend only on their distance. We know that this is the case with respect to the attraction of bodies at a distance, but we cannot make any such assumption concerning the internal forces of bodies of whose internal constitution we know next to nothing.

We cannot even assert that all energy must be either potential or kinetic, though we may not be able to conceive any other form. Nevertheless, the principle has been demonstrated by dynamical reasoning to be absolutely true for systems fulfilling certain conditions, and it has been proved by experiment to be true within the limits of error of observation, in cases where the energy takes the forms of heat, magnetisation, electrification, &c., so that the following statement is one which, if we cannot absolutely affirm its necessary truth, is worthy of being carefully tested, and traced into all the conclusions which are implied in it.

GENERAL STATEMENT OF THE CONSERVATION OF ENERGY.

'*The total energy of any body or system of bodies is a quantity which can neither be increased nor diminished by any mutual action of these bodies, though it may be transformed into any of the forms of which energy is susceptible.'*

If by the application of mechanical force, heat, or any other kind of action to a body, or system of bodies, it is made to pass through any series of changes, and at last to return in all respects to its original state, then the energy communicated to the system during this cycle of operations must be equal to the energy which the system communicates to other bodies during the cycle.

For the system is in all respects the same at the beginning and at the end of the cycle, and in particular it has the same amount of energy in it; and therefore, since no internal action of the system can either produce or destroy energy, the quantity of energy which enters the system must be equal to that which leaves it during the cycle.

The reason for believing heat not to be a substance is that it can be generated, so that the quantity of it may be increased to any extent, and it can also be destroyed, though this operation requires certain conditions to be fulfilled.

The reason for believing heat to be a form of energy is that heat may be generated by the application of work, and that for every unit of heat which is generated a certain quantity of mechanical energy disappears. Besides, work may be done by the action of heat, and for every foot-pound of work so done a certain quantity of heat is put out of existence.

Now when the appearance of one thing is strictly connected with the disappearance of another, so that the amount which exists of the one thing depends on and can be calculated from the amount of the other which has disappeared, we conclude that the one has been formed at the expense of the other, and that they are both forms of the same thing.

Hence we conclude that heat is energy in a peculiar form. The reasons for believing heat as it exists in a hot body to be in the form of kinetic energy—that is, that the particles of the hot body are in actual though invisible motion—will be discussed afterwards.

CHAPTER V.

ON THE MEASUREMENT OF PRESSURE AND OTHER INTERNAL FORCES, AND OF THE EFFECTS WHICH THEY PRODUCE.

EVERY force acts between two bodies or parts of bodies. If we are considering a particular body or system of bodies, then those forces which act between bodies belonging to this system and bodies not belonging to the system are called External Forces, and those which act between the different parts of the system itself are called Internal Forces.

If we now suppose the system to be divided in imagination into two parts, we may consider the forces external to one of the parts to be, first, those which act between that part and bodies external to the system, and, second, those which act between the two parts of the system. The combined effect of these forces is known by the actual motion or rest of the part to which they are applied, so that, if we know the resultant of the external forces on each part, we can find that of the internal forces acting between the two parts.

Thus, if we consider a pillar supporting a statue, and imagine the pillar divided into two parts by a horizontal plane at any distance from the ground, the internal force between the two parts of the pillar may be found by considering the weight of the statue and that part of the pillar which is above the plane. The lower part of the pillar presses on the upper part with a force which exactly counterbalances this weight. This force is called a Pressure. In the same way we may find the internal force acting through any horizontal section of a rope which supports a heavy body to be a Tension equal to the weight of the heavy body and of the part of the rope below the imaginary section.

The internal force in the pillar is called Longitudinal Pressure, and that in the rope is called Longitudinal Tension. If this pressure or tension is uniform over the whole horizontal section, the amount of it per square inch can be found by dividing the whole force by the number of square inches in the section.

The internal forces in a body are called Stresses, and longitudinal pressure and tension are examples of particular kinds of stress. It is shown in treatises on Elasticity that the most general kind of stress at any point of a body may be represented by three longitudinal pressures or tensions in directions at right angles to each other.

For instance, a brick in a wall may support a vertical pressure depending on the height of the wall above it, and also a horizontal pressure in the direction of the length of the wall, depending on the thrust of an arch abutting against the wall, while in the direction perpendicular to the face of the wall the pressure is that of the atmosphere.

In solid bodies, such as a brick, these three pressures may be all independent, their magnitude being limited only by the strength of the solid, which will break down if the force applied to it exceeds a certain amount.

In fluids, the pressures in all directions must be equal, because the very slightest difference between the pressures in the three directions is sufficient to set the fluid in motion.

The subject of fluid pressure is so important to what follows that I think it worth while, at the risk of repeating what the student ought to know, to state what we mean by a fluid, and to show from the definition that the pressures in all directions are equal.

DEFINITION OF A FLUID.—*A fluid is a body the contiguous parts of which act on one another with a pressure which is perpendicular to the surface which separates those parts.*

Since the pressure is entirely perpendicular to the surface, there can be no friction between the parts of a fluid in contact.

Theorem.—The pressures in any two directions at a point of a fluid are equal. For, let the plane of the paper be that of the two given directions, and draw an isosceles triangle whose sides are perpendicular to the two directions respectively, and consider the equilibrium of a small triangular prism of which this triangle is the base. Let P Q be the pressures perpendicular to the sides, and R that perpendicular to the base. Then, since these three forces are in equilibrium, and since R makes equal angles with P and Q, P and Q must be equal. But the forces on which P and Q act are also equal; therefore the pressures referred to unit of area on these forces are equal, which was to be proved.

FIG. 8.

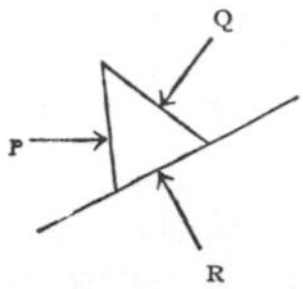

Q

P

R

A great many substances may be found which perfectly fulfil this definition of a fluid when they are at rest, and they are therefore called fluids. But no existing fluid fulfils the definition when it is in motion. In a fluid in motion the pressures at a point may be greater in one direction than in another, or, what is the same thing, the force between two parts may not be perpendicular to the surface which separates those parts.

If a fluid could be found which fulfilled the definition when in motion as well as when at rest, it would be called a Perfect Fluid. All actual fluids are imperfect, and exhibit the phenomenon of internal friction or viscosity, by which their motion after being stirred about in a vessel is gradually stopped, and the energy of the motion is converted into heat.

The degree of viscosity varies from that of tar to that of water, or ether, or hydrogen gas, but no actual fluid is perfect in the sense of the definition when in motion.

The pressure *at any point* of a fluid is the ratio of the whole pressure on a small surface to the area of that surface when the area of the surface is made to diminish indefinitely, but so that the centre of gravity of the surface always coincides with the given point.

This pressure is sometimes called hydrostatic pressure, to distinguish it from longitudinal pressure. Both kinds of pressure are measured by the number of units of force in the pressure on unit of area; for instance, in pounds' weight on the square inch or square foot, and in kilogrammes' weight on the square mètre. Both these measures are gravitation measures, and must be multiplied by the value of the intensity of gravity to reduce them to absolute measures.

Pressures are also measured in terms of the height of a column of water or of mercury, which would produce by its weight an equal pressure. Thus a pressure of 16 feet of water is nearly equal to 1,000 pounds' weight on the square foot, and a pressure of 4 inches of water is more nearly equal to 101 grains' weight on the square inch.

In the metrical system the pressure of water on a surface at any depth is expressed by the product of the depth into the area of the surface. If we employ the mètre as the measure of length, the pressure will be expressed in tonnes' weight, but if we use the decimètre, centimètre, or millimètre, the pressure will be expressed in kilogrammes, grammes, or milligrammes respectively, in gravitation measure.

The density of mercury at 0° C. is 13·596 times that of water at 4° C. Hence the pressure due to a given depth of mercury is about 13·6 times that of an equal depth of water.

The Barometer.—The pressure of the air is generally measured by means of the mercurial barometer. This barometer consists of a glass tube closed at one end and filled with mercury, from which all air and moisture are expelled by boiling it in the tube. The tube is then placed with its open end in a vessel of mercury, and its closed end raised till the tube is vertical. The mercury is found to stand at a certain level in the tube, the height of which above the level of the mercury in the vessel or cistern is called the height of the barometer.

The surface of the mercury in the cistern is exposed to the pressure of the air, while the surface of the mercury in the tube is exposed only to the pressure of whatever is in the tube above it. The only known substance which can be there is the vapour of mercury, the pressure of which at ordinary temperatures is so small that it may be neglected, so that the pressure of the air may be measured by that due to the difference of level of the mercury in the tube and in the cistern.

The pressure of the atmosphere is, as we know, very variable, and is different in different places; but for various purposes it is convenient to use, as a large unit of pressure, a pressure not very different from the average atmospheric pressure at the mean level of the sea. This unit of pressure is called an atmosphere, and is used in measuring pressures in steam-engines and boilers. Its exact value in the metrical system is the pressure due to a depth of 760 millimètres of mercury at 0° C. at Paris, where the force of gravity is 9·80868 mètres. This is equal to 1·033 kilogrammes' weight on the square centimètre. In absolute measure it is equal to 1,013,237, the gramme, the centimètre, and the second being the fundamental units.

In the British system an atmosphere is defined as the pressure due to a depth of 29·905 inches of mercury at 32° F. at London, where the force of gravity is 32·1889 feet, and is, roughly, $14\frac{3}{4}$ pounds' weight on the square inch. It is therefore 0·99968 of the atmosphere of the metrical system.

ON THE ALTERATION OF THE DIMENSIONS AND VOLUME OF BODIES BY MECHANICAL FORCES AND BY HEAT.

We have seen that effects of the same kind in changing the form or volume of bodies are produced by mechanical force and by heat. We cannot therefore fully understand the effects of heat alone on these bodies without at the same time considering those of mechanical force.

We have first to explain, from a purely geometrical point of view, the various kinds of change of form of which a body is capable, considering only those cases in which every part of the body undergoes a similar change of form. We shall use the word strain to express generally any alteration of form of a body.

Longitudinal Strain.—Suppose the body to be elongated or compressed in one direction only, so that if two points in the body lie in a line parallel to this direction, their distance will be increased or diminished in a certain ratio, but if the line joining the points be perpendicular to this direction the length of the line will not be altered.

This is called longitudinal extension or compression, or more generally longitudinal strain, and is measured by the fraction of its original length by which any longitudinal line in the body is elongated or contracted.

General Strain.—Such an alteration of the form of the body may take place simultaneously or successively in three directions at right angles to each other. This system of three longitudinal strains is shown in treatises on the motion of continuous bodies to be the most general kind of strain of which a body is capable.

We shall, however, only consider two cases in particular.

1st. *Isotropic Strain.*—When the strains in the three directions at right angles to each other are all equal, the form of the body remains similar to itself, and it expands or contracts equally in all directions, as most solid bodies do when heated.

Since each of the three longitudinal strains of which this strain is compounded increases or diminishes the volume by a fraction of itself equal to the value of the longitudinal strain, it follows that when each of the strains is a very small fraction, the total increment of volume is equal to the original volume multiplied by the sum of the three strains.

The ratio of the increment of volume to the original volume is called the cubical expansion when positive, or the cubical contraction when negative, and it appears, from what we have said, that when the strains are small the cubical expansion is equal to the sum of the longitudinal extensions, or, when these are equal, to three times the longitudinal extension.

2nd. *Shearing Strain.*—The other particular case is when the dimensions of the body are extended in one direction in the ratio of a to 1, and contracted in a perpendicular direction in the ratio of 1 to a. In this case there is no alteration of volume, but the body is distorted.

WORK DONE BY A STRESS ON A BODY WHOSE FORM IS CHANGING OR IS UNDERGOING A STRAIN.

We shall in the first place suppose that the stress continues constant during the change of form which we consider. If during a considerable change of form the stress undergoes considerable change, we may divide the whole operation into parts, during each of which we may regard the stress as constant, and find the total work by summation.

The general rule is that, if the stress and the strain are of the same type, the work done on unit of volume during any strain is the product of the strain into the average value of the stress.

If, however, the stress be of a type conjugate to the strain, no work is done.

Thus, if the stress be a longitudinal one, we must multiply the average value of the stress by the longitudinal strain in the same direction, and the result is not affected by the magnitude of the longitudinal strains in directions at right angles to the stress.

If the stress be a hydrostatic pressure, we must multiply the average value of this pressure by the cubical compression to find the work done on the body per unit of volume, and the result is not affected by any strain of distortion which does not change the volume of the body.

Hence the work done by external forces on a fluid when its volume is diminished is equal to the product of the average pressure into the diminution of volume, and if the fluid expands and overcomes the resistance of external forces, the work done by the fluid is measured by the product of the increase of volume, into the average pressure during that increase.

The consideration of the work gained or lost during the change of volume of a fluid is so important that we shall calculate it from the beginning.

WORK DONE BY A PISTON ON A FLUID.

Let us suppose that the fluid is in communication with a cylinder in which a piston is free to slide.

Fig. 9.

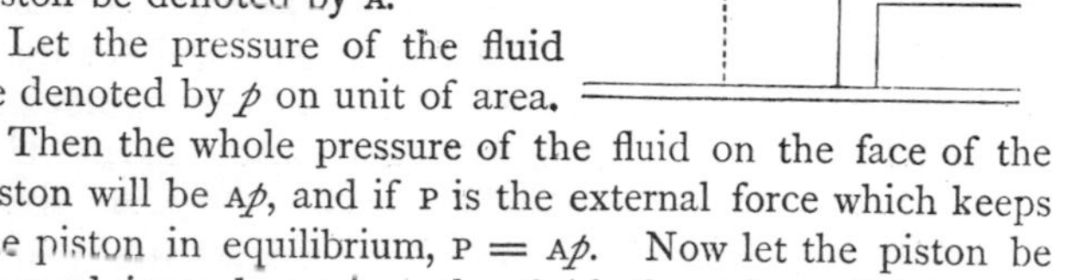

Let the area of the face of the piston be denoted by A.

Let the pressure of the fluid be denoted by p on unit of area.

Then the whole pressure of the fluid on the face of the piston will be Ap, and if P is the external force which keeps the piston in equilibrium, $P = Ap$. Now let the piston be pressed inwards against the fluid through a distance x. The volume of the cylinder occupied by the fluid will be diminished by a volume $V = Ax$, because the volume of a cylinder is equal to the area of its base multiplied by its height.

If the force P continues uniform, or if P is the average value of the external force during this motion, the work done by the external force will be $W = Px$.

If we put for P its value in terms of p, the pressure of the fluid per unit of area, this becomes

$$W = Apx;$$

and if we remember that Ax is equal to V, this becomes

$$W = Vp,$$

or the work done by the piston against the fluid is equal to the diminution of the volume of the fluid multiplied by the average value of the hydrostatic pressure.

It will be observed that this result is independent of the area of the piston, and of the form and capacity of the vessel with which the cylinder communicates.

If, for convenience, we suppose that the area of the piston is unity, then putting $A = 1$ we shall have $P = p$ and $V = x$, so that the linear distance travelled by the piston is numerically equal to the volume displaced.

ON INDICATOR DIAGRAMS.

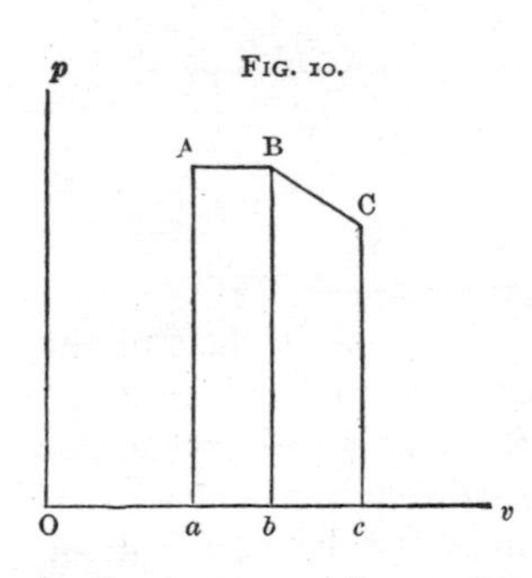

Fig. 10.

I shall now describe a method of studying the action of a fluid of variable volume, which was invented by James Watt, as a practical method of determining the work done by the steam-engine, and of which the construction has been gradually perfected, till it is now capable of tracing every part of the action of the steam in the most rapidly working engines.

At present, however, I shall use this method as a means of explaining and representing to the eye the working of a fluid. This use of the indicator diagram, which was introduced by Clapeyron, has been greatly developed by Rankine in his work on the steam-engine.

Let O v be a horizontal straight line, and O p a vertical line. On O v (which we shall call the line of volumes) take distances O a, O b, O c to represent the volume occupied by the fluid at different times, and at a b c erect perpendiculars a A, b B, c C, representing, on a convenient scale, the pressure of the fluid at these different times.

(For instance, we may suppose that, in the scale of volumes, one inch, measured horizontally, represents a volume equal to a cubic foot; and that in the scale of pressures, one inch, measured vertically, represents a pressure of 1,000 pounds' weight on the square foot.)

Let us now suppose that the volume increases from O *a* to O *b*, while the pressure remains constant, so that *a* A = *b* B.

Then the increase of volume is measured by *a b*, and the pressure which is overcome by the expansion of the fluid by *a* A or *b* B, so that the work done by the fluid is represented by the product of these quantities, or *a b*, *a* A, that is, the area of the rectangle A *a b* B.

On the scale which we have assumed, every square inch of the area of the figure A B *b a* represents 1,000 foot-pounds of work.

We have supposed the pressure to remain constant during the change of volume. If this is not the case, but if the pressure changes from *b* B to *c* C, while the volume changes from O *b* to O *c*, then if we take *b c* small enough, we may suppose the pressure to change uniformly from the one value to the other, so that we may take the mean value of the pressure to be $\frac{1}{2}$(B *b* + C *c*). Multiplying this by *b c*, we get $\frac{1}{2}$(B *b* + C *c*) *b c*, which is the well-known expression for the area of the strip B C *c b*, supposing B C a straight line.

The work done by the fluid is therefore still equal to the area enclosed by B C, the two vertical lines from its extremities, and the horizontal line O *v*.

In general, if the volume and pressure of the fluid are made to vary in any manner whatever, and if a point P be made at the same time to move so that its horizontal distance from the line O *p* represents the volume which the fluid occupies at that instant, while its vertical distance from O *v* represents the hydrostatic pressure of the fluid at the same instant, and if, at the beginning and end of the path traced by P, vertical lines be drawn to meet O *v*, then, if the path of P does not intersect itself, the area between these boundaries represents the work done by the fluid against external forces, if it lies on the right-hand side of the path of the tracing point. If the area lies on the left-hand side of the path, it represents the work done by the external forces on the fluid.

If the path of P returns into itself so as to form a loop or

FIG. 11.

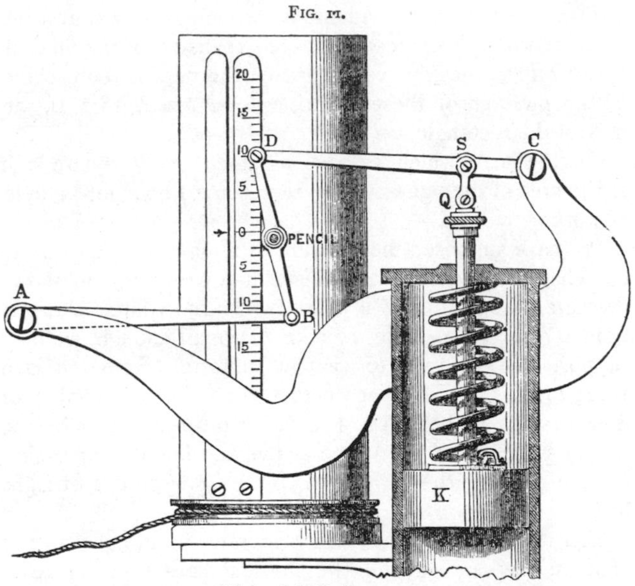

Richards's Indicator.

closed figure, then the vertical lines at the beginning and end of the path will coincide, so that it is unnecessary to draw them, and the work will be represented by the area of the loop itself. If P in its circuit goes round the loop in the direction of the hands of a watch, then the area represents the work done by the fluid against external forces; but if P goes round the loop in the opposite direction, the area of the loop represents the work done by the external forces on the fluid.

In the indicator as constructed by Watt and improved by McNaught and Richards, the steam or other fluid is put in connection with a small cylinder containing a piston. When the fluid presses this piston and raises it, the piston presses against a spiral spring, so constructed that the distance through which the spring is compressed is proportional to the pressure on the piston. In this way the height of the piston of the indicator is at all times a measure of the pressure of the fluid.

The piston also carries a pencil, the point of which presses lightly against a sheet of paper which is wrapped round a vertical cylinder capable of turning round its axis.

This cylinder is connected with the working piston of the engine, or with some part of the engine which moves along with the piston, in such a way that the angle through which the cylinder turns is always proportional to the distance through which the working piston has moved.

If the indicator is not connected with the steam pipe, the cylinder will turn beneath the point of the pencil, and a horizontal line will be drawn on the paper. This line corresponds to $o\,v$, and is called the line of no pressure.

But if the steam be admitted below the indicator piston, the pencil will move up and down, while the paper moves horizontally beneath it, and the combined motion will trace out a line on the paper, which is called an indicator diagram.

When the engine works regularly, so that each stroke is similar to the last, the pencil will trace out the same curve at every stroke, and by examining this curve we may learn much about the action of the engine. In particular, the area of the curve represents the amount of work done by the steam at each stroke of the engine.

If the indicator had been connected with a pump, in which the external forces do work on the fluid, the tracing point would move in the opposite direction round the diagram, and its area would indicate the amount of work done on the fluid during the stroke.

Hitherto we have confined our attention to the work done by the pressure on the piston, and have not been concerned with the cause of the alteration of volume of the fluid. The increase of volume may, for anything we know, arise from an additional supply being introduced into the cylinder, as when steam is introduced from the boiler, and the diminution of volume may arise from the escape of the fluid from the cylinder.

As we are now going to use the diagram for the purpose of explaining the properties of bodies when acted on by heat and by mechanical force, we shall suppose that the body, whether fluid or partly solid, is placed in a cylinder with one end closed, and that its volume is measured by the distance of the piston from the closed end of the cylinder.

FIG. 12.

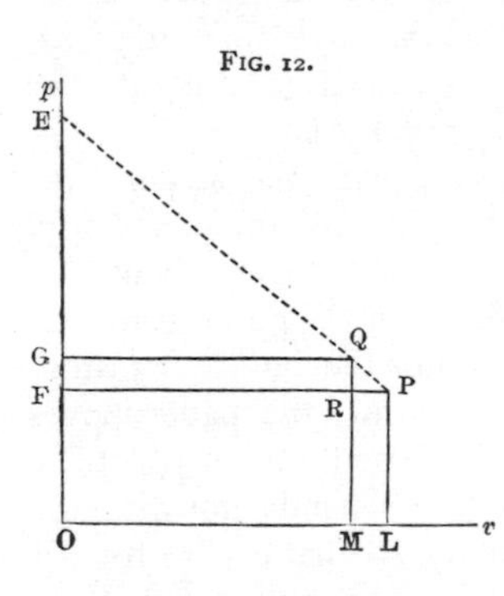

If at any instant the volume of the body is v and its pressure p, we represent this fact by means of the point P in the diagram, drawing O L along the line of volumes to represent v, and L P vertical to represent p.

In this way the position of a point in the diagram may be made to indicate the volume and the pressure of a body at any instant.

Now let the pressure be increased, the temperature remaining the same, then the volume of the fluid will be diminished. (It is manifest that an increase of pressure can never produce an increase of volume, for in that case the force would produce a motion in the contrary direction to that in which it acts, and we should have a source of inexhaustible energy.)

Let the pressure, therefore, increase from O F to O G, and let the consequent diminution of volume be from O L to O M, and complete the rectangle O G Q M.

Then the point P indicates the original and Q the final condition of the fluid with respect to pressure and volume, and all the intermediate states of the fluid will be represented by points in a line, straight or curved, which joins P and Q.

The work done by the pressure on the fluid is represented by the area of the figure P Q M L, which is on the left hand of the tracing point as it moves along P Q.

If P F and Q M intersect in R, then P R represents the actual diminution of volume, and R Q the actual increase of pressure. The actual volume is represented by F P, so that the cubical compression is represented by the ratio of P R to F P.

DEFINITION OF THE ELASTICITY OF A FLUID.—*The elasticity of a fluid under any given conditions is the ratio of any small increase of pressure to the cubical compression thereby produced.*

Since the cubical compression is a numerical quantity, the elasticity is a quantity of the same kind as a pressure.

To express the elasticity of the fluid by means of the diagram, join P Q by a straight line, and produce it till it meets the vertical line O p in E; then F E is a pressure equal to the elasticity of the fluid in the state represented by P, and under conditions which cause its state to vary in a manner represented by the line P Q.

For it is plain that F E is to R Q in the ratio of P F to P R,

$$\text{or } FE = \frac{RQ}{\frac{PR}{PF}} = \frac{\text{increment of pressure}}{\text{cubical compression}} = \text{elasticity.}$$

Hence if the relation between the volume and the pressure of a fluid under certain conditions, as for instance at a given temperature, is represented by a curve traced out by P, the elasticity of the fluid when in the state represented by P may be found by drawing P E a tangent to the curve at P, and P F a horizontal line. The portion F E of the vertical line O p cut off between these lines represents, on the scale of pressures, the elasticity of the fluid.

We have hitherto supposed that the temperature of the body remains the same during its compression from the volume P F to the volume Q G. This is the most common supposition when the elasticity of a fluid is to be measured. But in most bodies a compression produces a rise of temperature, and if the heat is not allowed to escape, the effect of this will be to make the increment of pressure greater than in the case of constant temperature. Hence every substance has two elasticities, one corresponding to constant temperature, and the other corresponding to the case where no heat is allowed to escape. The first value is applicable to stresses and strains which are long continued, so that the substance acquires the temperature of surrounding bodies. The second value is applicable to the case of rapidly changing forces, as in the case of the vibrations of bodies which produce sounds, in which there is not time for the temperature to be equalised by conduction. The elasticity in these cases is always greater than in the case of uniform temperature.

CHAPTER VI.

ON LINES OF EQUAL TEMPERATURE, OR ISOTHERMAL LINES ON THE INDICATOR DIAGRAM.

IF the pressure is made to vary while the temperature remains constant, the volume will diminish as the pressure increases, and the point P will trace out a line in the diagram which is called a line of equal temperature, or an isothermal line. By means of this line we can show the whole behaviour of the substance under various pressures at that particular temperature.

By making experiments on the substance at other temperatures, and drawing the isothermal lines belonging to these temperatures, we can express all the relations between the pressure, volume, and temperature of the substance.

In the diagram, each isothermal line should be marked with the temperature to which it corresponds in degrees, and the lines should be drawn for every degree, or for every ten or every hundred degrees, according to the purpose for which the diagram is intended.

When the volume and the pressure are known, the temperature is a determinate quantity, and it is easy to see how from any two of these three quantities we can determine the third. Thus if the curved lines in the diagram are the lines of equal temperature, the temperature corresponding to each being indicated by the numeral at the end of the line, we can solve three problems by means of this diagram.

1. Given the pressure and the volume, to find the temperature.

Lay off O L on the line of volumes to represent the given volume, and O F on the line of pressures to represent the given pressure, then draw F P horizontal and L P vertical, to determine the point P. If the point P falls on one of the lines of equal temperature, the numeral attached to that line indicates the temperature. If the point P falls between two of the lines, we must estimate its distance from the two nearest lines, and then as the sum of these distances is to the distance from the lower line of temperature, so is the difference of temperature of the two lines to the excess of the true temperature above that of the lower line.

2. Given the volume and temperature to find the pressure.

Lay off O L to represent the volume and draw L P vertical, and let P be the point where this line cuts the line of the given temperature. Then L P represents the required pressure.

3. Given the pressure and temperature, to find the volume.

FIG. 13.

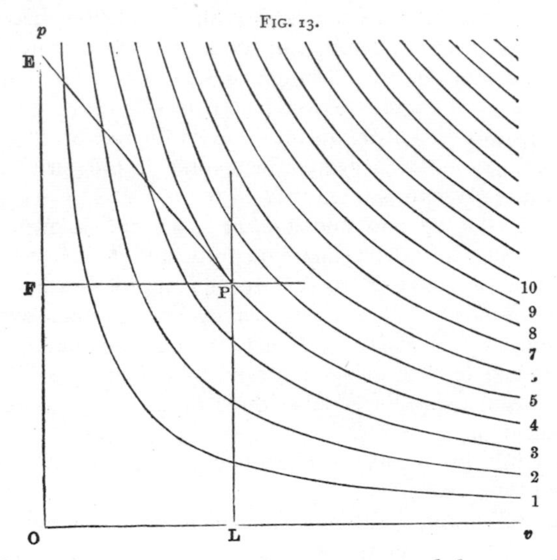

Lay off O F to represent the pressure and draw F P horizontal till it meets the line of the given temperature in P, then F P represents the required volume.

ON THE FORM OF THE ISOTHERMAL CURVES IN DIFFERENT CASES.

The Gaseous State.

If the substance is in the gaseous state, as common air is at every pressure and temperature to which we have been able to subject it, then it is easy to draw the isothermal curves by taking account of the laws of Boyle and Charles.

By Boyle's law the product of the volume and the pressure is always the same for the same temperature. Hence, in the curve, the area of the rectangle O L P F will be the same provided P be a point in the same isothermal curve.

The curve which has this property is known in geometry by the name of the rectangular hyperbola, the lines O v and O p being the asymptotes of the hyperbolas in fig. 13. The asymptotes are lines such that a point travelling along the curve in either direction continually approaches one or other of the asymptotes, but never reaches it. The physical interpretation of this is that if a gas fulfils Boyle's law, and if the temperature remain the same—

1. Suppose we travel along the curve in the direction leading toward O p, that is to say, suppose the pressure is gradually increased, then the volume will continually diminish, but always slower and slower; for, however much we increase the pressure, we can never reduce the volume to nothing, so that the isothermal line will never reach the line O p, though it continually approaches it. At the same time, if Boyle's law is fulfilled we can always, by doubling the pressure, reduce the volume to one half, so that by a sufficient increase of pressure the volume may be reduced till it is smaller than any prescribed quantity.

2. Suppose we travel in the other direction along the curve, that is to say, suppose we increase the volume of the vessel which contains the gas, then the point p approaches nearer and nearer to the line O v, but never actually reaches it. This shows that the gas will always expand so as to fill the vessel, and press upon it with a force represented by the distance from O v, and this pressure, though it diminishes as the vessel is enlarged, will never be reduced to nothing, however large the vessel may become.

Elasticity of a Perfect Gas.—Another property of the hyperbola is that if P E be drawn a tangent to the curve at P till it meets the asymptote, F E = O F. Now F E represents the elasticity of the substance, and O F the pressure. Hence the elasticity of a perfect gas is numerically equal to the pressure, when the temperature is supposed to remain constant during the compression.

The Liquid State.

In most liquids, the compression produced by the pressures which we are able to apply is exceedingly small. In the case of water, for example, under ordinary circumstances as to temperature, the application of a pressure equal to one atmosphere produces a compression of about 46 millionth parts of the volume, or 0·000046. Hence in drawing an indicator diagram for a liquid we must represent changes of volume on a much larger scale than in the case of gases, if the diagram is to have any visible features at all. The most convenient way is to suppose the line O L to represent, not the whole volume, but the excess of the volume above a thousand or a million of the units we employ.

It is manifest that the relation between the pressure and the volume of any substance must be such that no pressure, however great, can reduce the volume to nothing. Hence the isothermal lines cannot be straight lines, for a straight line, however slightly inclined to the line of no volumes O F, and however distant from it, must cut that line somewhere. The limited range of pressures which we are able to produce does not in some cases cause sufficient change of volume to indicate the curvature of the isothermal lines. We may suppose that for the small portion we are able to observe they are nearly straight lines.

The expansion due to an increase of temperature is also much smaller in the case of liquids than in the case of gases.

If, therefore, we were to draw the indicator diagram of a liquid on the same scale as that of a gas, the isothermal lines would consist of a number of lines very close together, nearly vertical, but very slightly inclined towards the line O F.

If, however, we retain the scale of pressures and greatly magnify the scale of volumes, the isothermal lines will be more inclined to the vertical and wider apart, but still very nearly straight lines. Liquids, however, which are near the critical point described at the end of this chapter are more compressible than even a gas.

The Solid State.

In solid bodies the compressibility and the expansion by heat are in general smaller than in liquids. Their indicator diagrams will therefore have the same general characteristics as those of liquids.

INDICATOR DIAGRAM OF A SUBSTANCE PART OF WHICH IS LIQUID AND PART VAPOUR.

Let us suppose that a pound of water is placed in a vessel and brought to a given temperature, say 212° F., and that by means of a piston the capacity of the vessel is made larger or smaller, the temperature remaining the same. If we suppose the vessel to be originally very large, say 100 cubic feet, and to be maintained at 212° F., then the whole of the water will be converted into steam, which will fill the vessel and will exert on it a pressure of about 575 pounds' weight on the square foot. If we now press down the piston, and so cause the capacity of the vessel to diminish, the pressure will increase nearly in the same proportion as the volume diminishes, so that the product of the numbers representing the pressure and volume will be nearly constant. When, however, the volume is considerably diminished, this product begins to diminish, that is to say, the pressure does not increase so fast as it ought to do by Boyle's law if the steam were a perfect gas. In the diagram, fig. 14, p. 114, the relations between the pressure and volume of steam at 212° are indicated by the curve *a b*. The pressure in atmospheres is marked on the right hand of the diagram, and the volume of one pound, in cubic feet, at the bottom.

When the volume is diminished to 26·36 cubic feet the

FIG. 14

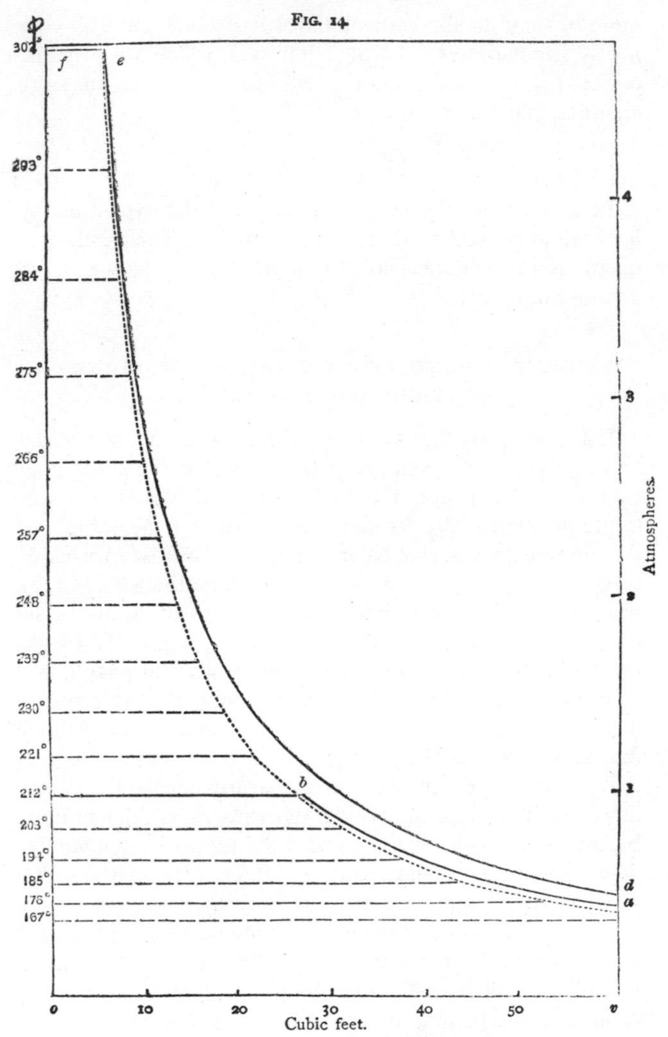

Isothermals for Steam and Water.

pressure is 2,116 lb., so that the product of the volume and pressure, instead of 57,500, is now reduced to 55,770. This departure from the law of Boyle, though not very large, is quite decided. The pressure and volume of the steam in this state are indicated by the point *b* in the diagram.

If we now diminish the volume and still maintain the same temperature, the pressure will no longer increase, but part of the steam will be converted into water; and as the volume continues to diminish, more and more of the steam will be condensed into the liquid form, while the pressure remains exactly the same, namely, 2,116 pounds' weight on the square foot, or one atmosphere. This is indicated by the horizontal line *b c* in the diagram.

This pressure will continue the same till all the steam is condensed into water at 212°, the volume of which will be 0·016 of a cubic foot, a quantity too small to be represented clearly in the diagram.

As soon as the volume, therefore, is reduced to this value there will be no more steam to condense, and any further reduction of volume is resisted by the elasticity of water, which, as we have seen, is very large compared with that of a gas.

We are now able to trace the isothermal line for water corresponding to the temperature 212°. When V is very great the curve is nearly of the form of an hyperbola for which $V P = 57{,}500$. As V diminishes, the curve falls slightly below the hyperbola, so that when $V = 26{\cdot}36$, $V P = 55{,}770$. Here, however, the line suddenly and completely alters its character, and becomes the horizontal straight line *b c*, for which $P = 2{,}116$, and this straight line extends from $V = 26{\cdot}36$ to $V = 0{\cdot}016$, when another equally sudden change takes place, and the line, from being exactly horizontal, becomes nearly but not quite vertical, nearly in the direction *c p*, for the pressure must be increased beyond the limits of our experimental methods long before any very considerable change is made in the volume of the water.

The isothermal line in a case of this kind consists of three parts. In the first part, *ab*, it resembles the isothermal lines of a perfect gas, but as the volume diminishes the pressure begins to be somewhat less than it should be by Boyle's law. This however, is only when the line approaches the second part of its course, *bc*, in which it is accurately horizontal. This part corresponds to a state in which the substance exists partly in the liquid and partly in the gaseous state, and it extends from the volume of the gas to the volume of the liquid at the same temperature and pressure. The third part of the isothermal line is that corresponding to the liquid state of the substance, and it may be considered as a line which on the scale of our diagrams would be very nearly vertical, and so near to the line *c p* that it cannot be distinguished from it.

In the diagram, fig. 14, the isothermal line of water for the temperature 212° F., the ordinary boiling point, is represented by *a b c p*, and that for 302° F. by *d e f p*.

At the temperature of 302° F. the pressure at which condensation takes place is much greater, being 9,966 pounds' weight on the square foot; and the volume to which the steam is reduced before condensation begins is much smaller, being 6·153 cubic feet. This is indicated by the point *e*. At this point the product V P is 61,321, which is considerably less than 65,209, its value when the volume is very great.

At this point condensation begins and goes on till the whole steam is condensed into water at 302° F., the volume of which is 0·0166 cubic feet. This volume is somewhat greater than the volume of the same water at 212° F.

It appears, therefore, that as the temperature rises the pressure at which condensation occurs is greater. It also appears that the diminution of volume when condensation takes place is less than at low temperatures, and this for two reasons. The first is, that the steam must be reduced to a smaller volume before condensation begins; and the second is, that the volume of the liquid when condensed is greater.

The dotted line in the diagram indicates the pressures and the volumes at which condensation begins at the various temperatures marked on the horizontal parts of the isothermal lines.

When the pressure and volume are those indicated by points above or on the right hand of this curve the whole substance is in the gaseous state. We may call this line the steam line. It is not an isothermal line.

If the scale of the diagram had been large enough to have represented the volume of the condensed water, we should have had another dotted line near the line o p, such that for points on the left hand of this line the whole substance is in the liquid state. We may call this the water line. For conditions of pressure and volume indicated by points between the two dotted lines, the substance is partly in the liquid and partly in the gaseous state. If we draw a horizontal line through the given point till it meets the two dotted lines, then the weight of steam is to the weight of water as the segment between the point and the water line is to the segment between the point and the steam line. In the lower part of the diagram for carbonic acid, fig. 15, p. 120, the isothermal lines are seen to consist of a curved portion on the right hand representing the gaseous state, a horizontal portion representing the process of condensation, and a nearly vertical portion representing the liquid state. The right-hand branch of the dotted line, which we must here call the gas line, corresponds to the steam line; and the left-hand branch, or liquid line, corresponds to the water line, which was not distinguishable in fig. 14.

Since these two lines, which we have called the steam line and the water line, continually approach each other as the temperature is raised, the question naturally arises, Do they ever meet? The peculiarity of the conditions indicated by points between these lines is that the liquid and its vapour can exist together under the same conditions as to temperature and pressure without the vapour being liquefied or the liquid evaporated. Outside of this region the substance must be either all vapour or all liquid.

If the two lines meet, then at the pressure indicated by the point of meeting there is no temperature at which the substance can exist partly as a liquid and partly as a vapour, but the substance must either be entirely converted from the state of vapour into the state of liquid at once and without condensation, or, since in this case the liquid and the vapour have the same density, it may be suspected that the distinctions we have been accustomed to draw between liquids and vapours have lost their meaning.

The answer to this question has been to a great extent supplied by a series of very interesting researches.

In 1822 M. Cagniard de la Tour [1] observed the effect of a high temperature upon liquids enclosed in glass tubes of a capacity not much greater than that of the liquid itself. He found that when the temperature was raised to a certain point, the substance, which till then was partly liquid and partly gaseous, suddenly became uniform in appearance throughout, without any visible surface of separation, or any evidence that the substance in the tube was partly in one state and partly in another.

He concluded that at this temperature the whole became gaseous. The true conclusion, as Dr. Andrews has shown, is that the properties of the liquid and those of the vapour continually approach to similarity, and that, above a certain temperature, the properties of the liquid are not separated from those of the vapour by any apparent distinction between them.

In 1823, the year following the researches of Cagniard de la Tour, Faraday succeeded in liquefying several bodies hitherto known only in the gaseous form, by pressure alone,

[1] *Annales de Chimie*, 2me série, xxi. et xxii.

and in 1826 he greatly extended our knowledge of the effects of temperature and pressure on gases. He considers that above a certain temperature, which, in the language of Dr. Andrews, we may call the critical temperature for the substance, no amount of pressure will produce the phenomenon which we call condensation, and he supposes that the temperature of 166° F. below zero is probably above the critical temperature for oxygen, hydrogen, and nitrogen.

Dr. Andrews has examined carbonic acid under varied conditions of temperature and pressure, in order to ascertain the relations of the liquid and gaseous states, and has arrived at the conclusion that the gaseous and liquid states are only widely separated forms of the same condition of matter, and may be made to pass one into the other without any interruption or breach of continuity.[1]

Carbonic acid is a substance which at ordinary temperatures and pressures is known as a gas. The measurements of Regnault and others show that as the pressure increases the volume diminishes faster than that of a gas which obeys the law of Boyle, and that as the temperature rises the expansion is greater than that assigned by the law of Charles.

The isothermal lines of the diagram of carbonic acid at ordinary temperatures and pressures are therefore somewhat flatter and also somewhat wider apart than those of the more perfect gases.

The diagram (p. 120) for carbonic acid is taken from Dr. Andrews's paper, with the exception of the dotted line showing the region within which the substance can exist as a liquid in presence of its vapour. The base line of the diagram corresponds, not to zero pressure, but to a pressure of 47 atmospheres.

The lowest of the isothermal lines is that of 13°·1 C. or 55°·6 F.

This line shows that at a pressure of about 47 atmospheres condensation occurs. The substance is seen to become

[1] *Phil. Trans.* 1869, p. 575.

FIG. 15.

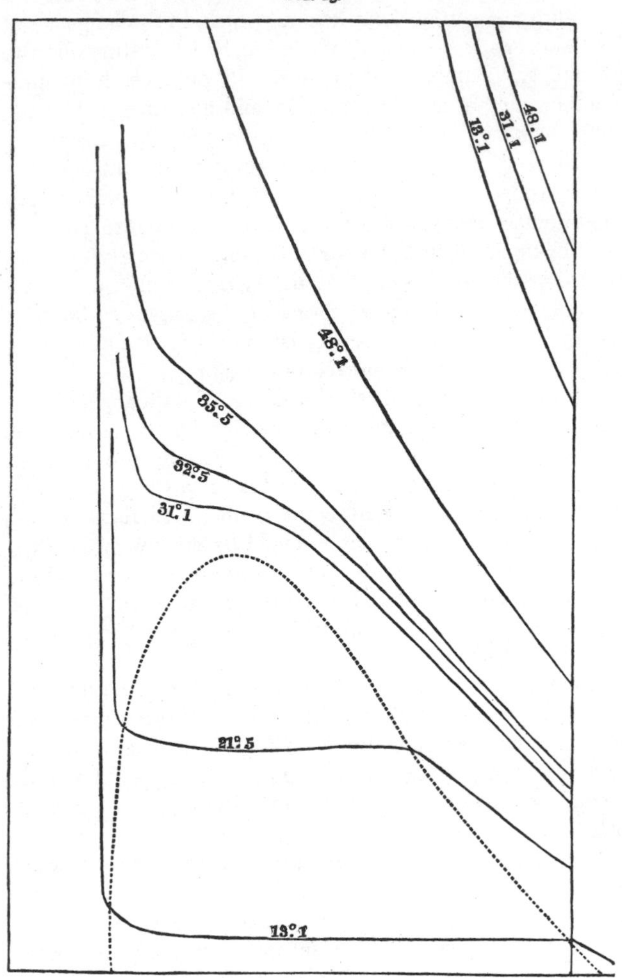

Isothermals of Carbonic Acid.

separated into two distinct portions, the upper portion being in the state of vapour or gas, and the lower in the state of liquid. The upper surface of the liquid can be distinctly seen, and where this surface is close to the sides of the glass containing the substance it is seen to be curved, as the surface of water is in small tubes.

As the volume is diminished, more of the substance is liquefied, till at last the whole is compressed into the liquid form.

I have described this isothermal line at greater length, that the student may compare the properties of carbonic acid at 55°·6 F. with those of water at 212° F.

1. The steam before condensation begins has properties agreeing nearly, though not quite, with those of a perfect gas. In carbonic acid the volume just before liquefaction commences is little more than three-fifths of that of a perfect gas at the same temperature and pressure. The corresponding isothermal lines for air are given in the diagram, and it will be seen how much the carbonic acid isothermal has fallen below that of air before liquefaction begins.

2. The steam when condensed into water occupies less than the sixteen-hundredth part of the volume of the steam. The liquid carbonic acid, on the other hand, occupies nearly a fifth part of its volume just before condensation. We are therefore able to draw the dotted line of complete condensation in this diagram, though in the case of water it would have required a microscope to distinguish it from the line of no volume.

3. The steam when condensed into water at 212° has properties not differing greatly from those of cold water. Its dilatability by heat and its compressibility by pressure are probably somewhat greater than when cold, but not enough to be noticed when the measurements are not very precise.

Liquid carbonic acid, as was first observed by Thilorier, dilates as the temperature rises to a greater degree than even a gas, and, as Dr. Andrews has shown, it yields to pressure much more than any ordinary liquid. From Dr. Andrews's experiments it also appears that its compressibility diminishes as the pressure increases. These results are apparent even in the diagram. It is, therefore, far more compressible than any ordinary liquid, and it appears from the experiments of Andrews that its compressibility diminishes as the volume is reduced.

It appears, therefore, that the behaviour of liquid carbonic acid under the action of heat and pressure is very different from that of ordinary liquids, and in some respects approaches to that of a gas.

If we examine the next of the isothermals of the diagram, that for 21°·5 C. or 70°·7 F., the approximation between the liquid and the gaseous states is still more apparent. Here condensation takes place at about 60 atmospheres of pressure, and the liquid occupies nearly a third of the volume of the gas. The exceedingly dense gas is approaching in its properties to the exceedingly light liquid. Still there is a distinct separation between the gaseous and liquid states, though we are approaching the critical temperature. This critical temperature has been determined by Dr. Andrews to be 30°·92 C. or 87°·7 F. At this temperature, and at a pressure of from 73 to 75 atmospheres, carbonic acid appears to be in the critical condition. No separation into liquid and vapour can be detected, but at the same time very small variations of pressure or of temperature produce such great variations of density that flickering movements are observed in the tube 'resembling in an exaggerated form the appearances exhibited during the mixture of liquids of different densities, or when columns of heated air ascend through colder strata.'

The isothermal line for 31°·1 C. or 88° F. passes above this critical point. During the whole compression the substance is never in two distinct conditions in different parts of the tube. When the pressure is less than 73 atmospheres the isothermal line, though greatly flatter than that of a perfect gas, resembles it in general features. From 73 to 75 atmospheres the volume diminishes very rapidly, but by no means suddenly, and above this pressure the volume diminishes more gradually than in the case of a perfect gas, but still more rapidly than in most liquids.

In the isothermals for 32°·5 C. or 90°·5 F. and for 35°·5 C. or 95°·9 F. we can still observe a slight increase of compressibility near the same part of the diagram, but in the isothermal line for 48° 1 C. or 118°·6 F. the curve is concave upwards throughout its whole course, and differs from the corresponding isothermal line for a perfect gas only by being somewhat flatter, showing that for all ordinary pressures the volume is somewhat less than that assigned by Boyle's law.

Still at the temperature of 118°·6 F. carbonic acid has all the properties of a gas, and the effects of heat and pressure on it differ from their effects on a perfect gas only by quantities requiring careful experiments to detect them.

We have no reason to believe that any phenomenon similar to condensation would occur, however great a pressure were applied to carbonic acid at this temperature.

In fact, by a proper management we can convert carbonic acid gas into a liquid without any sudden change of state.

If we begin with carbonic acid gas at 50° F. we may first heat it till its temperature is above 88° F., the critical point. We then gradually increase the pressure to, say, 100 atmospheres. During this process no sign of liquefaction occurs. Finally we cool the substance, still under the pressure of 100 atmospheres, to 50° F. During this process no sudden change of state can be observed, but carbonic acid at 50° F. and under a pressure of 100 atmospheres has all the properties of a liquid. At the temperature of 50° F. we cannot convert carbonic acid gas into a liquid without a sudden condensation, but by this process, in which the pressure is applied at a high temperature, we have caused the substance to pass from an undoubtedly gaseous to an undoubtedly liquid state without at any time undergoing an abrupt change similar to ordinary liquefaction.

I have described the experiments of Dr. Andrews on carbonic acid at greater length because they furnish the most complete view hitherto given of the relation between the liquid and the gaseous state, and of the mode in which the properties of a gas may be continuously and imperceptibly changed into those of a liquid.

The critical temperatures of most ordinary liquids are much higher than that of carbonic acid, and their pressure in the critical state is very great, so that experiments on the critical state of ordinary liquids are difficult and dangerous. M. Cagniard de la Tour estimated the temperature and pressure of the critical state to be:

	Temperature Fahr.	Pressure (Atmospheres)
Ether	369°·5	37·5
Alcohol	497°·5	119·0
Bisulphide of Carbon	504°·5	66·5
Water	773°·0	—

In the case of water the critical temperature was so high that the water began to dissolve the glass tube which contained it.

The critical temperature of what are called the permanent gases is probably exceedingly low, so that we cannot by any known method produce a degree of cold sufficient, even when applied along with enormous pressure, to condense them into the liquid state.

It has been suggested by Professor James Thomson[1] that the isothermal curves for temperatures below the critical temperature are only apparently, and not really, discontinuous, and that their true form is somewhat similar in its general features to the curve A B C D E F G H K.

The peculiarity of this curve is, that between the pressures

[1] *Proceedings of the Royal Society*, 1871, No. 130.

indicated by the horizontal lines B F and D H, any horizontal line such as C E G cuts the curve in three different points. The literal interpretation of this geometrical circumstance would be that the fluid at this pressure, and at the temperature of the isothermal line, is capable of existing in three different states. One of these, indicated by C, evidently corresponds to the liquid state. Another, indicated by G, corresponds to the gaseous state. At the intermediate point E the slope of the curve indicates that the volume and the pressure increase and diminish together.

FIG. 16.

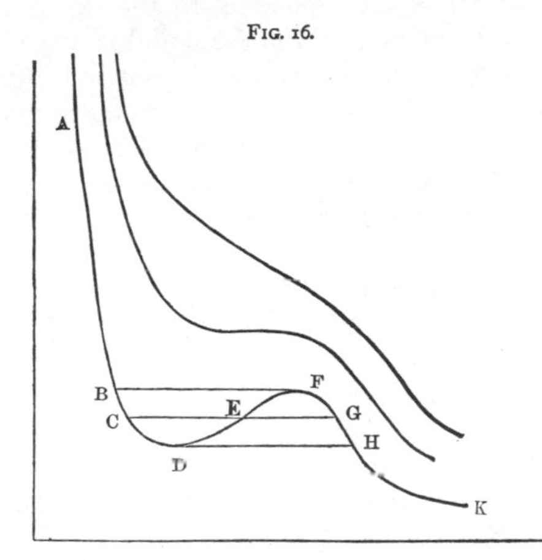

No substance having this property can exist in stable equilibrium, for the very slightest disturbance would make it rush into the liquid or the gaseous state. We may therefore confine our attention to the points C and G.

According to the theory of exchanges, as explained at p. 303, when the liquid is in contact with its vapour the rate of evaporation depends on the temperature of the liquid, and the rate of condensation on the density of the vapour. Hence for every temperature there is a determinate vapour-density, and therefore a determinate pressure, represented by the horizontal line C G, at which the evaporation exactly balances the condensation. At the pressure indicated by this horizontal line the liquid will be in equilibrium with its vapour. At all greater pressures the vapour, if in contact with the liquid, will be condensed; and at all smaller pressures the liquid, if in contact with its vapour, will evaporate. Hence the isothermal line, as deduced from experiments of the ordinary kind, will consist of the curve A B C, the straight line C G, and the curve G K.

But it has been pointed out by Prof. J. Thomson that by suitable contrivances we may detect the existence of other parts of the isothermal curve. We know that the portion of the curve corresponding to the liquid state extends beyond the point C; for if the liquid is carefully freed from air and other impurities, and is not in contact with anything but the sides of a vessel to which it closely adheres, the pressure may be reduced considerably below that indicated by the point C, till at last, at some point between C and D, the phenomenon of *boiling with bumping* commences, as described at p. 25.

Let us next consider the substance wholly in the state of vapour, as indicated by the point K, and let it be kept at the same temperature and gradually compressed till it is in the state indicated by the point G. If there are any drops of the liquid in the vessel, or if the vessel affords any facilities for condensation, condensation will now begin. But if there are no facilities for condensation, the pressure may be increased and the volume diminished till the state of the vapour is that which is represented by the point F. At this point condensation must take place if it has not begun before.[1]

The existence of this variability in the circumstances of condensation, though seemingly probable, is not as yet established by experiment, like that of the variability in the circumstances of evaporation. Prof. J. Thomson suggests that by investigating the condensation produced by the rapid expansion of vapour in a vessel provided with a

[1] See the chapter on Capillarity.

steam-jacket, the existence of this part of the isothermal curve might be established.

The state of things, however, represented by the portion of the isothermal curve D E F, can never be realised in a homogeneous mass, for the substance is then in an essentially unstable condition, since the pressure increases with the volume. We cannot, therefore, expect any experimental evidence of the existence of this part of the curve, unless, as Prof. J. Thomson suggests, this state of things may exist in some part of the thin superficial stratum of transition from a liquid to its own gas, in which the phenomena of capillarity take place.

CHAPTER VII.

ON THE PROPERTIES OF A SUBSTANCE WHEN HEAT IS PREVENTED FROM ENTERING OR LEAVING IT.

HITHERTO we have considered the properties of substance only with respect to the volume occupied by a pound of the substance, the pressure acting on every square foot or inch, and the temperature of the substance, which we have assumed to be uniform. We suppose the temperature measured by a thermometer, and when, in order to change the state of the body, heat must be supplied to it or taken from it, we have supposed this to be done without paying any attention to the quantity of heat required in each case. For the actual measurements of such quantities of heat we must employ the processes described in our chapter on Calorimetry, or others equivalent to them. Before entering on these considerations, however, we shall examine the very important case in which the changes which take place are effected without any passage of heat either into the substance from without or out of the substance into other bodies.

For the sake of associating the statement of scientific facts with mental images which are easily formed, and which preserve the statements in a form always ready for use, we shall suppose that the substance is contained in a cylinder fitted with a piston, and that both the cylinder and the piston are absolutely impermeable to heat, so that not only is heat prevented from getting out or in by passing completely through the cylinder or piston, but no heat can pass between the enclosed substance and the matter of the cylinder or piston itself.

No substance in nature is absolutely impermeable to heat, so that the image we have formed can never be fully realised; but it is always possible to ascertain, in each particular case, that heat has not entered or left the substance, though the methods by which this is done and the arrangements by which the condition is fulfilled are complicated. In the present discussion it would only distract our attention from the most important facts to describe the details of physical experiments. We therefore reserve any description of actual experimental methods till we can explain them in connexion with the principles on which they are founded. In explaining these principles we make use of the most suitable illustrations, without assuming that they are physically possible.

We therefore suppose the substance placed in a cylinder, and its volume and pressure regulated and measured by a piston, and we suppose that during the changes of volume and pressure of the substance no heat either enters it or leaves it.

In order to represent the relation between the volume and the pressure, we suppose a curve traced on the indicator diagram during the motion of the piston, exactly as in the case of the isothermal lines formerly described. The only difference is that whereas in the case of the isothermal lines the substance was maintained always at one and the same temperature, in the present case no heat is allowed to enter or leave the substance, which, as we shall see, is a condition of quite a different kind.

The line drawn on the indicator diagram in the latter case has been named by Professor Rankine an Adiabatic line, because it is defined by the condition that heat is not allowed to pass through (διαβαίνειν) the vessel which confines the substance.

Since the properties of the substance under this condition are completely defined by its adiabatic lines, it will assist us in understanding these properties if we associate them with the corresponding features of the adiabatic lines.

The first thing to be observed is that as the volume diminishes the pressure invariably increases. In fact, if under any circumstances the pressure were to diminish as the volume diminishes, the substance would be in an unstable state, and would either collapse or explode till it attained a condition in which the pressure increased as the volume diminished.

Hence the adiabatic lines slope downwards from left to right in the indicator diagram as we have drawn it.

If the pressure be continually increased, up to the greatest pressure which we can produce, the volume continually diminishes, but always slower and slower, so that we cannot tell whether there is or is not a limiting volume such that no pressure, however great, can compress the substance to a smaller volume.

We cannot, in fact, trace the lines upward beyond a certain distance, and therefore we cannot assert anything of the upper part of their course, except that they cannot recede from the line of pressures, because in that case the volume would increase on account of an increase of pressure.

If, on the other hand, we suppose the piston to be drawn out so as to allow the volume to increase, the pressure will diminish.

If the substance is in the gaseous form, or assumes that form during the process, the substance will continue to exert pressure on the piston even though the volume is enormously increased, and we have no experimental reason to believe that the pressure would be reduced to nothing, however much the volume were increased. For gaseous bodies, therefore, the lines extend indefinitely in the direction of the line of volumes, continually approaching but never reaching it.

With respect to substances which are not originally in the gaseous form, some of them, when the pressure is sufficiently diminished, are known to assume that form, and it is plausibly argued that we have no evidence that any substance, however solid and however cold, if entirely free from external pressure, would not sooner or later become dissipated through space by a kind of evaporation.

The smell by which such metals as iron and copper may be recognised is adduced as an indication that bodies, apparently very fixed, are continually throwing off portions of themselves in some very attenuated form, and if in these cases we have no means of detecting the effluvium except by the smell, in other cases we may be deprived of this evidence by the circumstance that the effluvium does not affect our sense of smell at all.

Be this as it may, there are many substances the pressure of which seems to cease entirely when the volume has reached a certain value. Beyond this the pressure, if it exists, is far too small to be measured. The lines of such substances may without sensible error be considered as meeting the line of volumes within the limits of the diagram.

FIG. 17.

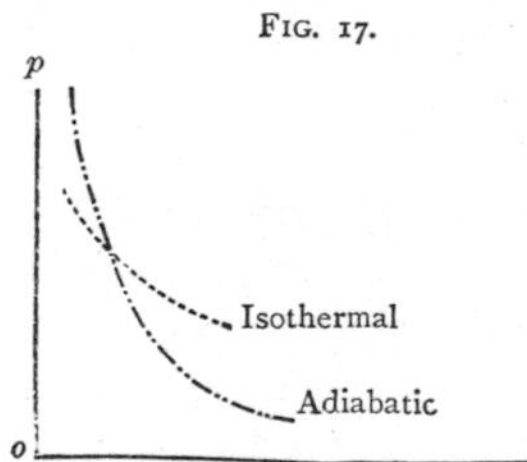

The next thing to be observed about the adiabatic lines is that where they cross the isothermal lines they are always inclined at a greater angle to the horizontal line than the isothermal lines.

In other words, to diminish the volume of a substance by a given amount requires a greater increase of pressure when the substance is prevented from gaining or losing heat than when it is kept at a constant temperature.

This is an illustration of the general principle that when the state of a body is changed in any way by the application of force in any form, and if in one case the body is subjected to some constraint, while in another case it is free from this constraint but similarly circumstanced in all other respects, then if during the change the body takes advantage of this freedom, less force will be required to produce the change than when the body is subjected to constraint.

In the case before us we may suppose the condition of constant temperature to be obtained by making the cylinder of a substance which is a perfect conductor of heat, and surrounding it with a very large bath of a fluid which is also a perfect conductor of heat, and which has so great a capacity for heat that all the heat it receives from or gives off to the substance in the cylinder does not sensibly alter its temperature.

The cylinder in this case is capable of constraining the substance itself, because it cannot get through the sides of the cylinder; but it is not capable of constraining the heat of the substance, which can pass freely out or in through the walls of the cylinder.

If we now suppose the walls of the cylinder to become perfect non-conductors of heat, everything remains the same, except that the heat is no longer free to pass into or out of the cylinder.

If in the first case the motion of the piston gives rise to any motion of the heat through the walls, then in the second case, when this motion is prevented, more force will be required to produce a given motion of the cylinder on account of the greater constraint of the system on which the force acts.

From this we may deduce the effect which the compression of a substance has on its temperature when heat is prevented from entering or leaving the substance.

We have seen that in every case the pressure increases more than it does when the temperature remains constant, or if the increase of pressure be supposed given, the diminution

Fig. 18.

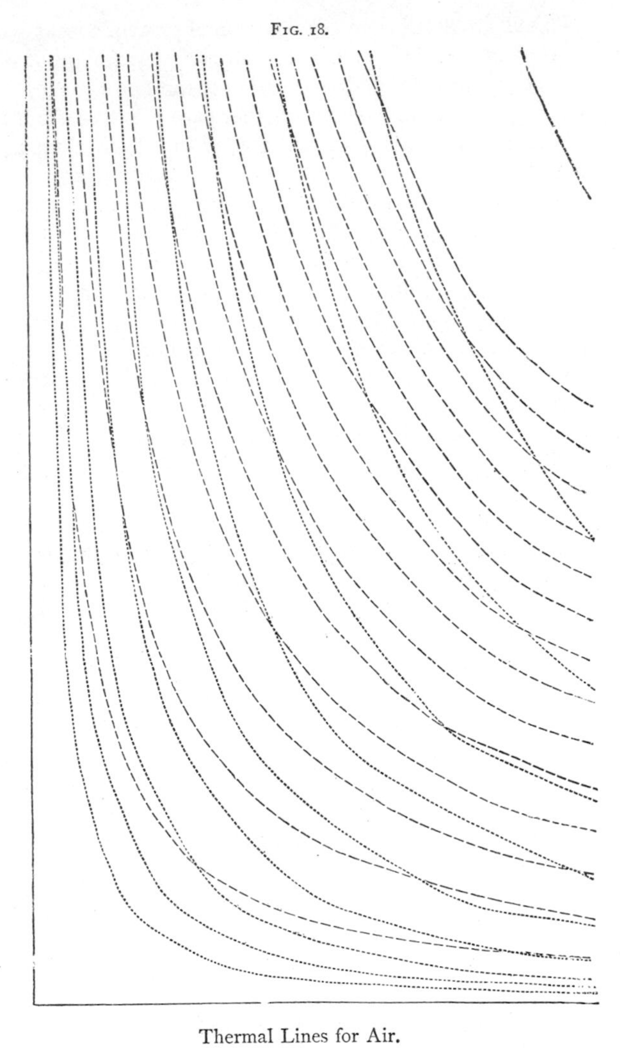

Thermal Lines for Air.

Isothermals — — — — — —
Adiabatics - - - - - - - - - - - -

of volume is less when the heat is confined. Hence the volume after the pressure is applied is greater when the heat is confined than when the temperature is constant.

Far the greater number of substances expand when their temperature is raised, so that for the same pressure a greater volume corresponds to a higher temperature. In these substances, therefore, compression produces a rise of temperature if heat is not allowed to escape; but if the walls of the cylinder permit the passage of heat, as soon as the temperature has begun to rise heat will begin to flow out, so that if the compression is effected slowly the principal thermal effect of the compression will be to make the substance part with some of its heat. The isothermal and adiabatic lines of air are given in fig. 18, p. 132. The adiabatic lines are more inclined to the horizontal than the isothermal lines.

There are, however, certain substances which contract instead of expanding when their temperature is raised. When pressure is applied to these substances the compression produced is, as in the former case, less when heat is prevented from passing than when the temperature is maintained constant. The volume after the application of pressure is therefore, as before, greater than when the temperature is constant; but since in these substances an increase of volume indicates a fall of temperature, it follows that, instead of being heated, they are cooled by compression, and that, if the walls of the cylinder permit the passage of heat, heat will flow in from without to restore the equilibrium of temperature.

During a change of state, when, at a given pressure, the volume alters considerably without change of temperature, as successive portions of the substance pass from the one state to the other, the isothermal lines are, as we have already remarked, horizontal. The adiabatic lines, however, are inclined downwards from left to right. Any increase of pressure will cause a portion of the substance to pass into that one of the two states in which its volume is least. In so doing it will give out heat if, as in the case of a liquid and its vapour, the substance gives out heat in passing into the denser state; but if, as in the case of ice and water, the ice requires heat to melt it into the denser form of water, then an increase of pressure will cause some of the ice to melt, and the mixture will become colder.

The isothermal and adiabatic lines for steam in presence of water are given in fig. 19, p. 135. The isothermal lines are here horizontal.

The steam line v V, which indicates the volume of one pound of saturated steam, is also drawn on the diagram. Its inclination to the horizontal line is less than that of the adiabatic lines. Hence when no heat is allowed to escape, an increase of pressure causes some of the water to become steam, and a diminution of pressure causes some of the steam to be condensed into water. This was first shown by Clausius and Rankine.

By means of diagrams of the isothermal and adiabatic lines the thermal properties of a substance can be completely defined, as we shall show in the subsequent chapters As a scientific method, this mode of representing the properties of the substance is by far the best, but in order to interpret the diagrams, some knowledge of thermodynamics is required. As a mere aid to the student in remembering the properties of a substance, the following mode of tracing the changes of volume and temperature at a constant pressure may be found useful, though it is quite destitute of those scientific merits which render the indicator diagrams so valuable in the investigation of physical phenomena.

The diagram on p. 137 represents the effect of the application of heat to a pound of ice at 0° F. The quantity of heat applied to the ice is indicated by the distance measured along the base line marked 'units of heat.' The volume of the substance is indicated by the length of the perpendicular from the base line cut off by the 'line of volume,' and the temperature is indicated by the length cut off by the dotted 'line of temperature.'

The specific heat of ice is about 0·5, so that it requires 16 units of heat to raise its temperature from 0° F. to 32° F. The specific gravity of ice at 32° F. is, according to Bunsen,

FIG. 19.

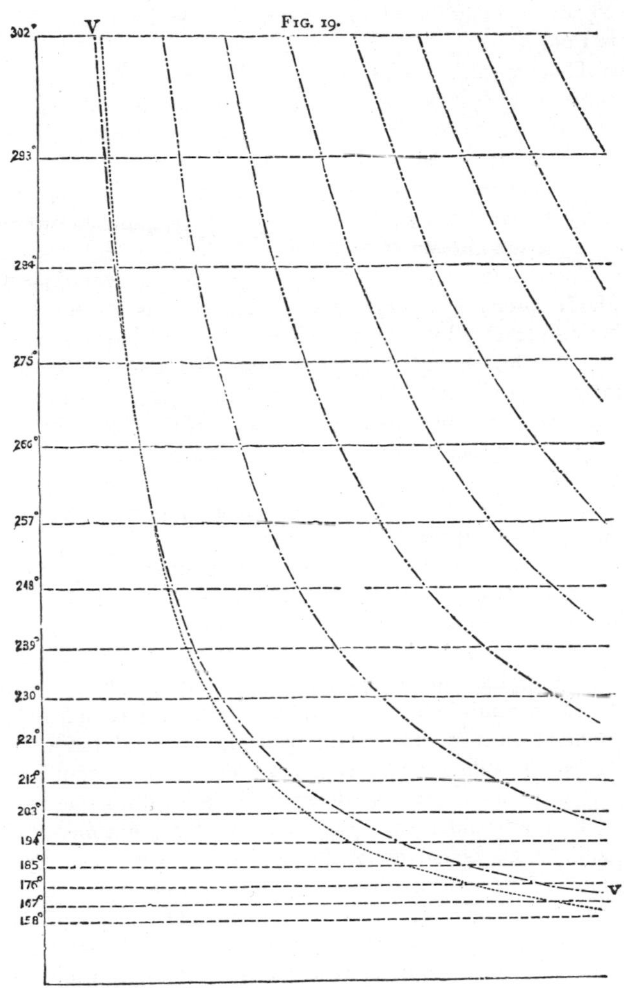

Thermal Lines of Steam and Water.

Isothermals — — — — — —
Adiabatics — · · — · · — · · —
Steam Line V V

0·91674, so that its volume, as compared with water at 39°·1, is 1·0908.

The ice now begins to melt, the temperature remains constant at 32° F., but the volume of ice diminishes and the volume of water increases, as is represented by the line marked 'volume of ice.' The latent heat of ice is 144° F., so that the process of melting goes on till 144 units of heat have been applied to the substance, and the whole is converted into water at 32° F.

The volume of the water at 32° F. is, according to M. Despretz, 1·000127. Its specific heat is at this temperature a very little greater than unity; it is exactly unity at 39°·1 F., and as the temperature rises the specific heat increases, so that to heat the water from 32° F. to 212° F. requires 182 units instead of 180. The volume of the water diminishes as the temperature rises from 32° F. to 39°·1 F., where it is exactly 1. It then expands, slowly at first, but more rapidly as the temperature rises, till at 212° F the volume of the water is 1·04315.

If we continue to apply heat to the water, the pressure being still that of the atmosphere, the water begins to boil. For every unit of heat, one nine hundred and sixty-fifth part of the pound of water is boiled away and is converted into steam, the volume of which is about 1,700 times that of the water from which it was formed. The diagram might be extended on a larger sheet of paper to represent the whole process of boiling the water away. This process would require 965 units of heat, so that the whole length of the base line from 0 would be 11·07 inches. At this point the water would be all boiled away, and the steam would occupy a volume of 1,700 times that of the water. The vertical line on the diagram which would represent the volume of the steam would be 3,400 inches, or more than 286 feet long. The temperature would be still 212° F. If we continue to apply heat to the steam, still at the atmospheric pressure, its temperature will rise in a perfectly uniform manner at

FIG. 20.

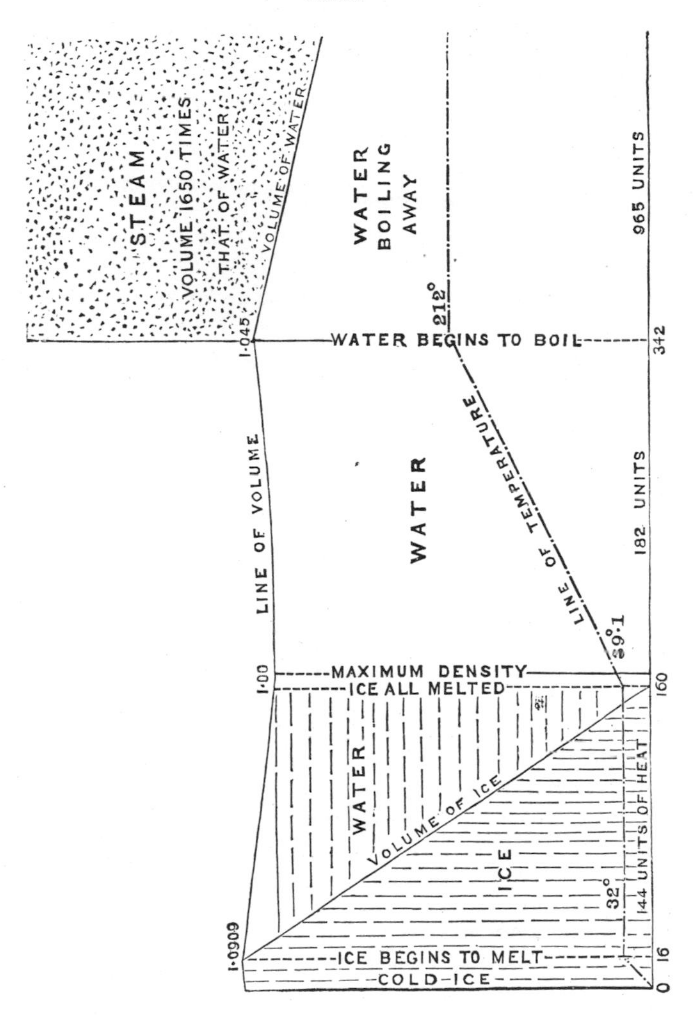
STEAM
VOLUME 1650 TIMES
THAT OF WATER
VOLUME OF WATER
WATER
BOILING
AWAY
965 UNITS
212°
1·045
WATER BEGINS TO BOIL
342
LINE OF VOLUME
WATER
LINE OF TEMPERATURE
182 UNITS
39°.1
1·00
MAXIMUM DENSITY
ICE ALL MELTED
160
WATER
VOLUME OF ICE
ICE
32°
144 UNITS OF HEAT
1·0909
ICE BEGINS TO MELT
16
COLD ICE
0

the rate of 2°·08 degrees for every unit of heat, the specific heat of steam being 0·4805.

The volume of the superheated steam also increases in a regular manner, being proportional to its absolute temperature reckoned from −960° F.

CHAPTER VIII.

ON HEAT ENGINES.

HITHERTO the only use we have made of the indicator diagram is to explain the relation between the volume and the pressure of a substance placed in certain thermal conditions. The condition that the temperature is constant gave us the isothermal lines, and the condition that no communication of heat takes place gave us the adiabatic lines. We have now to consider the application of the same method to the measurements of quantities of heat and quantities of mechanical work.

At p. 102 it was shown that if the pencil of the indicator moves from B to C, this shows that the volume of the substance has increased from O *b* to O *c*, under a pressure which was originally B *b* and finally C *c*.

The work done by the pressure of the substance against the piston during this motion is represented by the area B *c* C *b*, and since the volume *increases* during the process, it is the substance which does the work on the piston, and not the piston which does the work on the substance.

In heat engines of ordinary construction, such as steam engines and air engines, the form of the path described by the pencil depends on the mechanical arrangements of the engine, such as the opening and shutting of the valves which admit or carry off the steam.

For the purposes of scientific illustration, and for obtaining clear views of the dynamical theory of heat, we shall describe the working of an engine of a species entirely imaginary—one which it is impossible to construct, but very easy to understand.

This engine was invented and described by Sadi Carnot, in his 'Réflexions sur la Puissance motrice du Feu,' published in 1824. It is called Carnot's Reversible Engine for reasons which we shall explain.

All the arrangements connected with this engine are contrived for the sake of being explained, and are not intended to represent anything in the working of real engines.

Carnot himself was a believer in the material nature of heat, and was in consequence led to an erroneous statement of the quantities of heat which must enter and leave the engine. As our object is to understand the theory of heat, and not to give an historical account of the theory, we shall avail ourselves of the important step which Carnot made, while we avoid the error into which he fell.

FIG. 21.

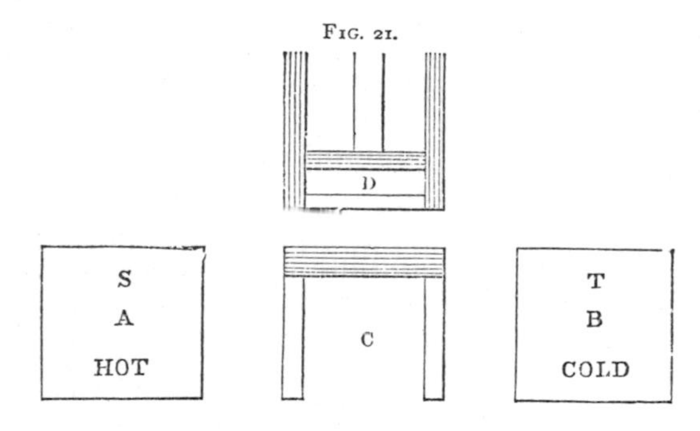

Let D be the working substance, which may be any substance whatever which is in any way affected by heat, but, for the sake of precision, we shall suppose it to be either air or steam, or partly steam and partly condensed water at the same temperature.

The working substance is contained in a cylinder fitted with a piston. The walls of the cylinder and the piston are supposed to be perfect non-conductors of heat, but the bottom of the cylinder is a perfect conductor of heat, and has so small a capacity for heat that the amount of heat required to raise its temperature may be left out of account. All the communication of heat between the working substance and things outside the cylinder is supposed to take place through this conducting bottom, and the quantities of heat are supposed to be measured as they pass through.

A and B are two bodies the temperatures of which are maintained uniform. A is kept always hot, at a temperature S, and B is kept always cold, at a temperature T. C is a stand to set the cylinder on, the upper surface of which is a perfect non-conductor of heat.

Let us suppose that the working substance is at the temperature T of the cold body B, and that its volume and pressure are represented in the indicator diagram by O a and a A, the point A being on the isothermal line A D corresponding to the lower temperature T.

Fig. 22.

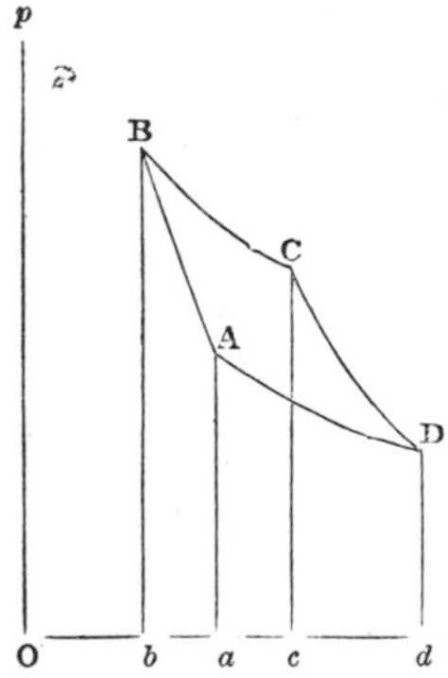

First Operation.—We now place the cylinder on the non-conducting stand C, so that no heat can escape, and we then force the piston down, so as to diminish the volume of the substance. As no heat can escape, the temperature rises, and the relation between volume and pressure at any instant will be expressed by the pencil tracing the adiabatic line A B.

We continue this process till the temperature has risen to S, that of the hot body A. During this process we have expended an amount of work on the substance which is represented by the area A B b a. If work is reckoned negative when it is spent on the substance, we must regard that employed in this first operation as negative.

Second Operation.—We now transfer the cylinder to the hot body A, and allow the piston gradually to rise. The immediate effect of the expansion of the substance is to make its temperature fall, but as soon as the temperature begins to fall, heat flows in from the hot body A through the perfectly conducting bottom, and keeps the temperature from falling below the temperature S.

The substance will therefore expand at the temperature S, and the pencil will trace out the line B C, which is part of the isothermal line corresponding to the upper temperature S.

During this process the substance is doing work by its pressure on the piston. The amount of this work is represented by the area B C *c b*, and it is to be reckoned positive.

At the same time a certain amount of heat, which we shall call H, has passed from the hot body A into the working substance.

Third Operation.—The cylinder is now transferred from the hot body A to the non-conducting body C, and the piston is allowed to rise. The indicating pencil will trace out the adiabatic line C D, since there is no communication of heat, and the temperature will fall during the process. When the temperature has fallen to T, that of the cold body, let the operation be stopped. The pencil will then have arrived at D, a point on the isothermal line for the lower temperature T.

The work done by the substance during this process is represented by the area C D *d c*, and is positive.

Fourth Operation.—The cylinder is placed on the cold body B. It has the same temperature as B, so that there is no transfer of heat. But as soon as we begin to press down the piston heat flows from the working substance into B, so that the temperature remains sensibly equal to T during the operation. The piston must be forced down till it has reached the point at which it was at the beginning of the first operation, and, since the temperature is also the same, the pressure will be the same as at first. The working substance, therefore, after these four operations, has returned exactly to its original state as regards volume, pressure, and temperature.

During the fourth operation, in which the pencil traces the portion D A of the isothermal line for the lower temperature, the piston does work on the substance, the amount of which is to be reckoned negative, and which is represented by the area D A *a d*.

At the same time a certain amount of heat, which we shall denote by *h*, has flowed from the working substance into the cold body B.

DEFINITION OF A CYCLE.—*A series of operations by which the substance is finally brought to the same state in all respects as at first is called a Cycle of operations.*

Total Work done during the Cycle.—When the piston is rising the substance is giving out work; this is the case in the second and third operations. When the piston is sinking it is performing work on the substance which is to be reckoned negative. Hence, to find the work performed by the substance we must subtract the area D A B *b d*, representing the negative work, from the positive work, B C D *d b*. The remainder, A B C D, represents the useful work performed by the substance during the cycle of operations. If we have any difficulty in understanding how this amount of work can be obtained in a useful form during the working of the engine, we have only to suppose that the piston when it rises is employed in lifting weights, and that a portion of the weight lifted is employed to force the piston down again. As the pressure of the substance is less when the piston is sinking than when it is rising, it is plain that the engine can raise a greater weight than that which is required to complete the cycle of operations, so that on the whole there is a balance of useful work.

Transference of Heat during the Cycle.—It is only in the second and fourth operations that there is any transfer of heat, for in the first and third the heat is confined by the non-conducting stand.

In the second operation a quantity of heat represented by H passes from the hot body A into the working substance at the upper temperature S, and in the fourth operation a quantity of heat represented by *h* passes from the working substance into the cold body B at the lower temperature T.

The working substance is left after the cycle of operations in precisely the same state as it was at first, so that the whole physical result of the cycle is—

1. A quantity, H, of heat taken from A at the temperature S.

2. The performance by the substance of a quantity of work represented by A B C D.

3. A quantity, *h*, of heat communicated to B at the temperature T.

APPLICATION OF THE PRINCIPLE OF THE CONSERVATION OF ENERGY.

It has long been thought by those who study natural forces that in all observed actions among bodies the work which is done is merely transferred from one body in which there is a store of energy into another, so as to increase the store of energy in the latter body.

The word energy is employed to denote the capacity which a body has of performing work, whether this capacity arises from the motion of the body, as in the case of a cannon-ball, which is able to batter down a wall before it can be stopped; or from its position, as in the case of the weight of a clock when wound up, which is able to keep the clock going for a week; or from any other cause, such as the elasticity of a watch-spring, the magnetisation of a compass needle, the chemical properties of an acid, or the heat of a hot body.

The doctrine of the conservation of energy asserts that all these different forms of energy can be measured in the same way that mechanical work is measured, and that if the whole energy of any system were measured in this way the mutual action of the parts of the system can neither increase nor diminish its total stock of energy.

Hence any increase or diminution of energy in a system must be traced to the action of bodies external to the system.

The belief in the doctrine of the conservation of energy has greatly assisted the progress of physical science, especially since 1840. The numerous investigations which have been made into the mechanical value of various forms of energy were all undertaken by men who believed that in so doing they were laying a foundation for a more accurate knowledge of physical actions considered as forms of energy. The fact that so many forms of energy can be measured on the hypothesis that they are all equivalent to mechanical energy, and that measurements conducted by different methods are consistent with each other, shows that the doctrine contains scientific truth.

To estimate its truth from a demonstrative point of view we must consider, as we have always to do in making such estimates, what is involved in a direct contradiction of the doctrine. If the doctrine is not true, then it is possible for the parts of a material system, by their mutual action alone, and without being themselves altered in any permanent way, either to do work on external bodies or to have work done on them by external bodies. Since we have supposed the system after a cycle of operations to be in exactly the same state as at first, we may suppose the cycle of operations to be repeated an indefinite number of times, and therefore the system is capable in the first case of doing an indefinite quantity of work without anything being supplied to it, and in the second of absorbing an indefinite quantity of work without showing any result.

That the doctrine of the conservation of energy is not self-evident is shown by the repeated attempts to discover a perpetual motive power, and though such attempts have been long considered hopeless by scientific men, these men themselves had repeatedly observed the apparent loss of energy in friction and other natural actions, without making any attempt or even showing any desire to ascertain what becomes of this energy.

The evidence, however, which we have of the doctrine is nearly if not quite as complete as that of the conservation of matter—the doctrine that in natural operations the quantity of matter in a system always remains the same though it may change its form.

No good evidence has been brought against either of these doctrines, and they are as certain as any other part of our knowledge of natural things.

The great merit of Carnot's method is that he arranges his operations in a cycle, so as to leave the working substance in precisely the same condition as he found it. We are therefore sure that the energy remaining in the working substance is the same in amount as at the beginning of the cycle. If this condition is not fulfilled, we should have to discover the energy required to change the substance from its original to its final state before we could make any assertion based upon the conservation of energy.

We have therefore got rid of the consideration of the energy residing in the working substance, which is called its *intrinsic energy*, and we have only to compare—

1. The original energy, which is a quantity H of heat at the temperature S of the hot body. This being communicated to the working substance, we get for the resulting energy—

2. A quantity of work done, represented by A B C D; and

3. A quantity *h* of heat at the temperature T of the cold body.

The principle of the conservation of energy tells us that the energy of the heat H at the temperature S exceeds that of the heat *h* at the temperature T by a quantity of mechanical energy represented by A B C D, which can be easily expressed in foot-pounds. This is admitted by all.

Now Carnot believed heat to be a material substance,

called caloric, which of course cannot be created or destroyed. He therefore concluded that, since the quantity of heat remaining in the substance is the same as at first, H, the quantity of heat communicated to it, and h, the quantity of heat abstracted from it, must be the same.

These two portions of heat, however, are, as Carnot observed, in different conditions, for H is at the temperature of the hot body, and h at that of the cold body, and Carnot concluded that the work of the engine was done at the expense of the fall of temperature, the energy of any distribution of heat being greater the hotter the body which contains it.

He illustrated this theory very clearly by the analogy of a water-mill. When water drives a mill the water which enters the mill leaves it again unchanged in quantity, but at a lower level. Comparing heat with water, we must compare heat at high temperature with water at a high level. Water tends to flow from high ground to low ground, just as heat tends to flow from hot bodies to cold ones. A water-mill makes use of this tendency of water, and a heat engine makes use of the corresponding property of heat.

The measurement of quantities of heat, especially when it has to be done in an engine at work, is an operation of great difficulty, and it was not till 1862 that it was shown experimentally by Hirn that h, the heat emitted, is really less than H, the heat received by the engine. But it is easy to see that the assumption that H is equal to h must be wrong.

For if we were to employ the engine in stirring a liquid, then the work A B C D spent in this way would generate an amount of heat which we may denote by $\mathfrak{H}$ in the liquid.

The heat H at the high temperature has therefore been used, and we find instead of it a quantity h at the low temperature, and also $\mathfrak{H}$ at the temperature of the liquid, whatever it is.

But if heat is material, and therefore $H = h$, then $h + \mathfrak{H}$ is greater than the original quantity H, and heat has been created, which is contrary to the hypothesis that it is material.

Besides this, we might have allowed the heat H to pass from the hot body to the cold body by conduction, either directly or through one or more conducting bodies, and in this case we know that the heat received by the cold body would be equal to the heat taken from the hot body, since conduction does not alter the quantity of heat. Hence in this case $H = h$, but no work is done during the transfer of heat. When, in addition to the transfer of heat, work is done by the engine, there ought to be some difference in the final result, but there will be no difference if h is still equal to H.

The hypothesis of caloric, or the theory that heat is a kind of matter, is rendered untenable, first by the proof given by Rumford, and more completely by Davy, that heat can be generated at the expense of mechanical work ; and, second, by the measurements of Hirn, which show that when heat does work in an engine, a portion of the heat disappears.

The determination of the mechanical equivalent of heat by Joule enables us to assert that the heat which is required to raise a pound of water from 39° F. to 40° F. is mechanically equivalent to 772 foot-pounds of work.

It is to be observed that in this statement nothing is said about the temperature of the body in which the heat exists. The heat which raises the pound of water from 39° F. to 40° F. may be taken from a vessel of cold water at 50° F., from a red-hot iron heater at 700° F., or from the sun at a temperature far above any experimental determination, and yet the heating effect of the heat is the same whatever be the source from which it flows. When heat is measured as a quantity, no regard whatever is paid to the temperature of the body in which the heat exists, any more than to the size, weight, or pressure of that body, just as when we determine the weight of a body we pay no attention to its other properties.

Hence if a body in a certain state, as to temperature, &c.,

is capable of heating so many pounds of water from 39° F. to 40° F. before it is itself cooled to a given temperature, say 40° F., then if that body, in its original state, is stirred about and its parts rubbed together so as to expend 772 foot-pounds of work in the process, it will be able to heat one pound more of water from 39° F. to 40° F. before it is cooled to the given temperature.

Carnot, therefore, was wrong in supposing that the mechanical energy of a given quantity of heat is greater when it exists in a hot body than when it exists in a cold body. We now know that its mechanical energy is exactly the same in both cases, although when in the hot body it is more available for the purpose of driving an engine.

In our statement of the four operations of Carnot's engine we arranged them so as to leave the result in a state in which we can interpret it either as Carnot did, or according to the dynamical theory of heat. Carnot himself began with the operation which we have placed second, the expansion at the upper temperature, and he directs us to continue the fourth operation, compression at the lower temperature, till exactly as much heat has left the substance as entered during the expansion at the upper temperature. The result of this operation would be, as we now know, to expel too much heat, so that after the substance had been compressed on the non-conducting stand to its original volume, its temperature and pressure would be too low. It is easy to amend the directions for the extent to which the outflow of heat is to be permitted, but it is still easier to avoid the difficulty by placing this operation last, as we have done.

We are now able to state precisely the relation between h, the quantity of heat which leaves the engine, and H, the quantity received by it. H is exactly equal to the sum of h, and the heat to which the mechanical work represented by A B C D is equivalent.

In all statements connected with the dynamical theory of heat it is exceedingly convenient to state quantities of heat in foot-pounds at once, instead of first expressing them in thermal units and then reducing the result to foot-pounds by means of Joule's equivalent of heat. In fact, the thermal unit depends for its definition on the choice of a standard substance to which heat is to be applied, on the choice of a standard quantity of that substance, and on the choice of the effect to be produced by the heat. According as we choose water or ice, the grain or the gramme, the Fahrenheit or the Centigrade scale of temperatures, we obtain different thermal units, all of which have been used in different important researches. By expressing quantities of heat in foot-pounds we avoid ambiguity, and, especially in theoretical reasonings about the working of engines, we save a great deal of useless phraseology.

As we have already shown how an area on the indicator diagram represents a quantity of work, we shall have no difficulty in understanding that it may also be taken to represent a quantity of heat equivalent to the same quantity of work, that is the same number of foot-pounds of heat.

We may therefore express the relation between H and h still more concisely thus:

The quantity, H, of heat taken into the engine at the upper temperature S exceeds the quantity, h, of heat given out by the engine at the lower temperature T by a quantity of heat represented by the area A B C D on the indicator diagram.

This quantity of heat is, as we have already shown, converted into mechanical work by the engine.

ON THE REVERSED ACTION OF CARNOT'S ENGINE.

The peculiarity of Carnot's engine is, that whether it is receiving heat from the hot body, or giving it out to the cold body, the temperature of the substance in the engine differs extremely little from that of the body in thermal communication with it. By supposing the conductivity of the bottom of the cylinder to be sufficiently great, or by supposing the motions of the piston to be sufficiently slow, we may make the actual difference of temperature which causes the flow of heat to take place as small as we please.

If we reverse the motion of the piston when the substance is in thermal communication with A or B, the first effect will be to alter the temperature of the working substance, but an exceedingly small alteration of temperature will be sufficient to reverse the flow of heat, if the motion is slow enough.

Now let us suppose the engine to be worked backwards by exactly reversing all the operations already described. Beginning at the lower temperature and volume O *a*, let it be placed on the cold body and expand from volume O *a* to O *d*. It will receive from the cold body a quantity of heat *h*. Then let it be compressed without losing heat to O *c*. It will then have the upper temperature S. Let it then be placed on the hot body and compressed to volume O *b*. It will give out a quantity of heat H to the hot body. Finally, let it be allowed to expand without receiving heat to volume O *a*, and it will return to its original state. The only difference between the direct and the reverse action of the engine is, that in the direct action the working substance must be a little cooler than A when it receives its heat, and a little warmer than B when it gives it out; whereas in the reverse action it must be warmer than A when it gives out heat, and cooler than B when it takes heat in. But by working the engine sufficiently slowly these differences may be reduced within any limits we please to assign, so that for theoretical purposes we may regard Carnot's engine as strictly reversible.

In the reverse action a quantity *h* of heat is taken from the cold body B, and a greater quantity H is given to the hot body A, this being done at the expense of a quantity of work measured by the area A D C B, which also measures the quantity of heat into which this work is transformed during the process.

The reverse action of Carnot's engine shows us that it is possible to transfer heat from a cold body to a hot one, but that this operation can only be done at the expense of a certain quantity of mechanical work.

The transference of heat from a hot body to a cold one may be effected by means of a heat engine, in which case part of it is converted into mechanical work, or it may take place by conduction, which goes on of itself, but without any conversion of heat into work. It appears, therefore, that heat may pass from hot bodies to cold ones in two different ways. One of these, in which a highly artificial engine is made use of, is nearly, but not quite completely, reversible, so that by spending the work we have gained, we can restore almost the whole of the heat from the cold body to the hot. The other mode of transfer, which takes place of itself whenever a hot and a cold body are brought near each other, appears to be irreversible, for heat never passes from a cold body to a hot one of itself, but only when the operation is effected by the artificial engine at the expense of mechanical work.

We now come to an important principle, which is entirely due to Carnot. If a given reversible engine, working between the upper temperature S and the lower temperature T, and receiving a quantity H of heat at the upper temperature, produces a quantity W of mechanical work, then no other engine, whatever be its construction, can produce a greater quantity of work, when supplied with the same amount of heat, and working between the same temperatures.

DEFINITION OF EFFICIENCY.—*If* H *is the supply of heat, and* W *the work done by an engine, both measured in foot-pounds, then the fraction* $\frac{W}{H}$ *is called the Efficiency of the engine.*

Carnot's principle, then, is that the efficiency of a reversible engine is the greatest that can be obtained with a given range of temperature.

For suppose a certain engine, M, has a greater efficiency between the temperatures S and T than a reversible engine N, then if we connect the two engines, so that M by its direct action drives N in the reverse direction, at each stroke of the compound engine N will take from the cold body B the heat h, and by the expenditure of work W give to the hot body A the heat H. The engine M will receive this heat H, and by hypothesis will do more work while transferring it to B than is required to drive the engine N. Hence at every stroke there will be an excess of useful work done by the combined engine.

We must not suppose, however, that this is a violation of the principle of conservation of energy, for if M does more work than N would do, it converts more heat into work in every stroke, and therefore M restores to the cold body a smaller quantity of heat than N takes from it. Hence, the legitimate conclusion from the hypothesis is, that the combined engine will, by its unaided action, convert the heat of the cold body B into mechanical work, and that this process may go on till all the heat in the system is converted into work.

This is manifestly contrary to experience, and therefore we must admit that no engine can have an efficiency greater than that of a reversible engine working between the same temperatures. But before we consider the results of Carnot's principle we must endeavour to express clearly the law which lies at the bottom of the reasoning.

The principle of the conservation of energy, when applied to heat, is commonly called the First Law of Thermodynamics. It may be stated thus : When work is transformed into heat, or heat into work, the quantity of work is mechanically equivalent to the quantity of heat.

The application of the law involves the existence of the mechanical equivalent of heat.

Carnot's principle is not deduced from this law, and indeed Carnot's own statement involved a violation of it. The law from which Carnot's principle is deduced has been called the Second Law of Thermodynamics.

Admitting heat to be a form of energy, the second law asserts that it is impossible, by the unaided action of natural processes, to transform any part of the heat of a body into mechanical work, except by allowing heat to pass from that body into another at a lower temperature. Clausius, who first stated the principle of Carnot in a manner consistent with the true theory of heat, expresses this law as follows:—

It is impossible for a self-acting machine, unaided by any external agency, to convey heat from one body to another at a higher temperature.

Thomson gives it a slightly different form:—

It is impossible, by means of inanimate material agency, to derive mechanical effect from any portion of matter by cooling it below the temperature of the coldest of the surrounding objects.

By comparing together these statements, the student will be able to make himself master of the fact which they embody, an acquisition which will be of much greater importance to him than any form of words on which a demonstration may be more or less compactly constructed.

Suppose that a body contains energy in the form of heat, what are the conditions under which this energy or any part of it may be removed from the body? If heat in a body consists in a motion of its parts, and if we were able to distinguish these parts, and to guide and control their motions by any kind of mechanism, then by arranging our apparatus so as to lay hold of every moving part of the body, we could, by a suitable train of mechanism, transfer the energy of the moving parts of the heated body to any other body in the form of ordinary motion. The heated body would thus be rendered perfectly cold, and all its thermal energy would be converted into the visible motion of some other body.

Now this supposition involves a direct contradiction to the second law of thermodynamics, but is consistent with the first law. The second law is therefore equivalent to a denial of our power to perform the operation just described, either by a train of mechanism, or by any other method yet discovered. Hence, if the heat of a body consists in the motion of its parts, the separate parts which move must be so small or so impalpable that we cannot in any way lay hold of them to stop them.

In fact, heat, in the form of heat, never passes out of a body except when it flows by conduction or radiation into a colder body.

There are several processes by which the temperature of a body may be lowered without removing heat from it, such as expansion, evaporation, and liquefaction, and certain chemical and electrical processes. Every one of these, however, is a reversible process, so that when the body is brought back by any series of operations to its original state, without any heat being allowed to enter or escape during the process, the temperature will be the same as before, in virtue of the reversal of the processes by which the temperature was lowered. But if, during the operations, heat has passed from hot parts of the system to cold by conduction, or if anything of the nature of friction has taken place, then to bring the system to its original state will require the expenditure of work, and the removal of heat.

We must now return to the important result demonstrated by Carnot, that a reversible engine, working between two given temperatures, and receiving at the higher temperature a given quantity of heat, performs at least as much work as any other engine whatever working under the same conditions. It follows from this that all reversible engines, whatever be the working substance employed, have the same efficiency, provided they work between the same temperature of the source of heat A and the same temperature of the refrigerator B.

Hence Carnot showed that if we choose two temperatures differing very slightly, say by $\frac{1}{1000}$ of a degree, the efficiency of an engine working between these temperatures will depend on the temperature only, and not on the substance employed, and this efficiency divided by the difference of temperatures is the quantity called *Carnot's function*, a quantity depending on the temperature only.

Carnot, of course, understood the temperature to be estimated in the ordinary way by means of a thermometer of a selected substance graduated according to one of the established scales, and his function is expressed in terms of the temperature so determined. But W. Thomson, in 1848, was the first to point out that Carnot's result leads to a method of defining temperature which is much more scientific than any of those derived from the behaviour of one selected substance or class of substances, and which is perfectly independent of the nature of the substance employed in defining it.

THOMSON'S ABSOLUTE SCALE OF TEMPERATURE.

Let T A B C represent the isothermal line corresponding to temperature T for a certain substance. For the sake of distinctness in the figure, I have supposed the substance to be partly in the liquid and partly in the gaseous state, so that the isothermal lines are horizontal, and easily distinguished from the adiabatic lines, which slope downwards to the right. The investigation, however, is quite independent of any such restriction as to the nature of the working substance. When the volume and pressure of the substance are those indicated by the point A, let heat be applied and let the substance expand, always at the temperature T, till a quantity of heat H has entered, and let the state of the substance be then indicated by the point B. Let the process go on till another equal quantity, H, of heat has entered, and let C indicate the resulting state. The process may be carried on so as to find any number of points on the isothermal line, such that for each point passed during the expansion of the substance a quantity H of heat has been communicated to it.

Now let A A′ A″, B B′ B″, C C′ C″ be adiabatic lines drawn through A B C, that is, lines representing the relation between volume and pressure when the substance is allowed to expand without receiving heat from without.

FIG. 23.

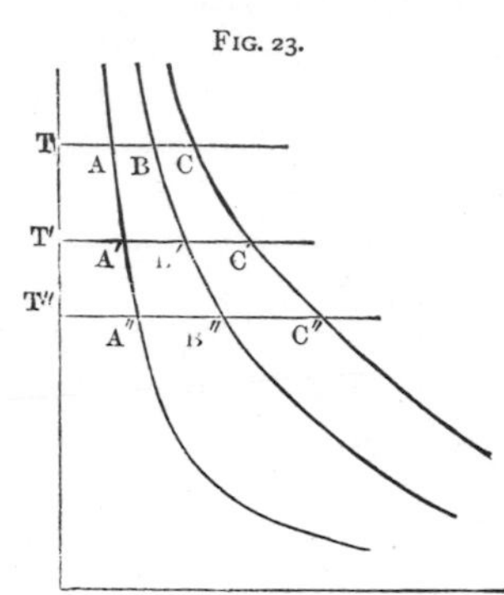

Let T′ A′ B′ C′ and T″ A″ B″ C″ be isothermal lines corresponding to the temperatures T′ and T″.

We have already followed Carnot's proof that in a reversible engine, working from the temperature T of the source of heat to the temperature T′ of the refrigerator, the work W produced by the quantity of heat H drawn from the source depends only on T and T′.

Hence, since A B and B C correspond to equal quantities of heat H received from the source, the areas A B B′ A′ and B C C′ B′, which represent the corresponding work performed, must be equal.

The same is true of the areas cut off by the adiabatic lines from the space between any other pair of isothermal lines.

Hence if a series of adiabatic lines be drawn so that the points at which they cut one of the isothermal lines correspond to successive equal additions of heat to the substance at that temperature, then this series of adiabatic lines will cut off a series of equal areas from the strip bounded by any two isothermal lines.

Now Thomson's method of graduating a scale of temperature is equivalent to choosing the points A A′ A″, from which to draw a series of isothermal lines, so that the area A B B′ A′ contained between two consecutive isothermals T and T′ shall be equal to the area A′ B′ B″ A″ contained between any other pair of consecutive isothermals T′ T″.

It is the same as saying that the number of degrees between the temperature T and the temperature T″ is to be reckoned proportional to the area A B B″ A″.

Of course two things remain arbitrary, the standard temperature which is to be reckoned zero, and the size of the degrees, and these may be chosen so that the absolute scale corresponds with one of the ordinary scales at the two standard temperatures, but as soon as these are determined the numerical measure of every other temperature is settled, in a manner independent of the laws of expansion of any one substance—by a method, in fact, which leads to the same result whatever be the substance employed.

It is true that the experiments and measurements required to graduate a thermometer on the principle here pointed out would be far more difficult than those required by the ordinary method described in the chapter on Thermometry. But we are not, in this chapter, describing convenient methods or good working engines. Our objects are intellectual, not practical, and when we have established theoretically the scientific advantages of this method of graduation, we shall be better able to understand the practical methods by which it can be realised.

We now draw the series of isothermal and adiabatic lines in the following way:

A particular isothermal line, that of temperature T, is cut by the adiabatic lines, so that the expansion of the substance between consecutive adiabatic lines corresponds to successive quantities of heat, each equal to H, applied to the substance. This determines the series of adiabatic lines.

The isothermal lines are drawn so that the successive isothermals cut off from the space between the pair of adiabatic lines A A′ A″ and B B′ B″ equal areas A B B′ A′, A′ B′ B″ A″, &c.

The isothermal lines so determined cut off equal areas

from every other pair of adiabatic lines, so that the two systems of lines are such that all the quadrilaterals formed by two pairs of consecutive lines are equal in area.

We have now graduated the isothermals on the diagram by a method founded on Carnot's principle alone, and independent of the nature of the working substance, and it is easy to see how by altering, if necessary, the interval between the lines and the line chosen for zero, we can make the graduation agree, at the two standard temperatures, with the ordinary scale.

EFFICIENCY OF A HEAT ENGINE.

Let us now consider the relation between the heat supplied to an engine and the work done by it as expressed in terms of the new scale of temperature.

If the temperature of the source of heat is T, and if H is the quantity of heat supplied to the engine at that temperature, then the work done by this heat depends entirely on the temperature of the refrigerator. Let T'' be the temperature of the refrigerator, then the work done by H is represented by the area $A\ B\ B''\ A''$, or, since all the areas between the isothermals and the adiabatics are equal, let $H\ C$ be the area of one of the quadrilaterals, then the work done by H will be $H\ C\ (T - T'')$. The quantity C depends only on the temperature T. It is called Carnot's Function of the temperature. We shall find a simple definition of it afterwards.

This, therefore, is a complete determination of the work done when the temperature of the source of heat is T. It depends only on Carnot's principle, and is true whether we admit the first law of thermodynamics or not.

If the temperature of the source is not T, but T', we must consider what quantity of heat is represented by the expansion $A'\ B'$ along the isothermal T'. Calling this quantity of heat H', the work done by an engine working between the temperatures T' and T'' is

$$W = H\ C\ (T' - T'').$$

Now Carnot supposed that $H' = H$, which would make the efficiency of the engine simply $\frac{W}{H} = C\,(T' - T'')$, where C is Carnot's function, a constant quantity on this supposition. But according to the dynamical theory of heat, we get by the first law of thermodynamics

$$H' = H - A\,B\,B'\,A',$$

the heat being measured as mechanical work, or

$$H' = H - H\,C\,(T - T').$$

On this theory, therefore, the efficiency of the engine working between T' and T'' is

$$\frac{W}{H'} = \frac{H\,C\,(T' - T'')}{H - H\,C\,(T - T')}$$

$$= \frac{T' - T''}{\frac{1}{C} + T' - T.}$$

ON ABSOLUTE TEMPERATURE.

We have now obtained a method of expressing differences of temperature in such a way that the difference of two temperatures may be compared with the difference of two other temperatures. But we are able to go a step farther than this, and to reckon temperature from a zero point defined on thermodynamic principles independently of the properties of a selected substance. We must carefully distinguish between what we are doing now on really scientific principles from what we did for the sake of convenience in describing the air thermometer. Absolute temperature on the air thermometer is merely a convenient expression of the laws of gases. The absolute temperature as now defined is independent of the nature of the thermometric substance. It so happens, however, that the difference between these two scales of temperature is very small. The reason of this will be explained afterwards.

It is plain that the work which a given quantity of heat H can perform in an engine can never be greater than the mechanical equivalent of that heat, though the colder the refrigerator the greater proportion of heat is converted into work. It is plain, therefore, that if we determine T'' the temperature of the refrigerator, so as to make W the work mechanically equivalent to H, the heat received by the engine, we shall obtain an expression for a state of things in which the engine would convert the whole heat into work, and no body can possibly be at a lower temperature than the value thus assigned to T''.

Putting $W = H'$, we find $T'' = T - \frac{1}{C}$.

This is the lowest temperature any body can have. Calling this temperature zero, we find

$$T = \frac{1}{C},$$

or the temperature reckoned from absolute zero is the reciprocal of Carnot's function C.

We have therefore arrived at a complete definition of the measure of temperature, in which nothing remains to be determined except the size of the degrees. Hitherto the size of the degrees has been chosen so as to be equal to the mean value of those of the ordinary scales. To convert the ordinary expressions into absolute temperatures we must add to the ordinary expression a constant number of degrees, which may be called the absolute temperature of the zero of the scale. There is also a correction varying at different parts of the scale, which is never very great when the temperature is measured by the air thermometer. We may now express the efficiency of a reversible heat engine in terms of the absolute temperature S of the source of heat, and the absolute temperature T of the refrigerator. If H is the quantity of heat supplied to the engine, and W is the quantity of work performed, both estimated in dynamical measure,

$$\frac{W}{H} = \frac{S - T}{S}.$$

The quantity of heat which is given out to the refrigerator at temperature T is $h = \text{H} - \text{W} = \text{H}\frac{\text{T}}{\text{S}}$, whence

$$\frac{\text{H}}{\text{S}} = \frac{h}{\text{T}} \text{ or } \frac{\text{H}}{h} = \frac{\text{S}}{\text{T}}$$

that is, in a reversible engine the heat received is to the heat rejected as the absolute temperature of the source is to the absolute temperature of the refrigerator.

This relation furnishes us with a method of determining the ratio of two temperatures on the absolute dynamical scale. It is independent of the nature of the substance employed in the reversible engine, and is therefore a perfect method considered from a theoretical point of view. The practical difficulties of fulfilling the required conditions and making the necessary measurements have not hitherto been overcome, so that the comparison of the dynamical scale of temperature with the ordinary scale must be made in a different way. (See p. 196.)

Let us now return to the diagram fig. 23 (p. 156), on which we have traced two systems of lines, the isothermals and the adiabatics. To draw an isothermal line through a given point requires only a series of experiments on the substance at a given temperature, as shown by a thermometer of any kind. To draw a series of these lines to represent successive degrees of temperature is equivalent to fixing a scale of temperature.

Such a scale might be defined in many different ways, each of which depends on the properties of some selected substance. For instance, the scale might be founded on the expansion of a particular substance at some standard pressure. In this case, if a horizontal line is drawn to represent the standard pressure, then the isothermal lines of the selected substance will cut this line at equal intervals. If, however, the nature of the substance or the standard pressure be different, the thermometric scale will be in general different. The scale might also be founded on the variation of pressure of a substance confined in a given space, as in the case of certain applications of the air thermometer.

It has also been proposed to define temperature so that equal increments of heat applied to a standard substance will produce equal increments of temperature. This method also fails to give results consistent for all substances, because the specific heats of different substances are not in the same ratio at different temperatures.

The only method which is certain to give consistent results, whatever be the substance employed, is that which is founded on Carnot's Function, and the most convenient form in which this method can be applied is that which defines the absolute temperature as the reciprocal of Carnot's Function. We shall see afterwards how a comparison can be made between the absolute temperature on the thermodynamic scale and the temperature as indicated by a thermometer of a particular kind of gas.

To draw an adiabatic line through any point requires only experiments on the substance. The series of adiabatic lines in our diagram is defined so that when the substance expands at a certain temperature T, the same quantity of heat H causes it to pass from one adiabatic line to the next. If we make this quantity of heat measured dynamically (that is in units of work) equal numerically to T, the absolute temperature of the standard isothermal line, then, as we have already shown, the area of every quadrilateral contained between two consecutive isothermals and two consecutive adiabatics will be $CH = \frac{T}{T} = 1$. To measure any area on the diagram we have only to count the number of these quadrilaterals contained in it. We must then mark the adiabatic lines, beginning with the line of no heat, with the indices 0, 1, 2, 3, &c., up to ϕ, the index or number of the line. This quantity, ϕ, is called by Rankine the Thermodynamic Function.

It is probably impossible to deprive a body entirely of

heat. If, however, we could do so, its temperature would be absolute zero, and the relation between its volume and its pressure at this temperature would be given by an isothermal line called the line of absolute cold. Of course we have no experimental data for determining the form of this line for any actual substance. We can only regard it as a limit beyond which no line in the diagram can extend.

In the seven chapters which follow we shall apply the principles of Thermodynamics to several important cases, avoiding as much as we can the introduction of the methods of the higher mathematics, by which most of the results are worked out in the original memoirs. It is hoped that the student may obtain some assistance from these chapters in his study of Thermodynamics. He may, however, especially when first reading the subject, pass on to Chapter XVI.

CHAPTER IX.

ON THE RELATIONS BETWEEN THE PHYSICAL PROPERTIES OF A SUBSTANCE.

LET $T_1 T_1$ and $T_2 T_2$ represent two isothermal lines corresponding to two consecutive degrees of temperature. Let $\phi_1 \phi_1$ and $\phi_2 \phi_2$ represent two consecutive adiabatic lines. Let A B C D be the quadrilateral which lies between both these pairs of lines. If the lines are drawn close enough to each other we may treat this quadrilateral as a parallelogram.

The area of this parallelogram is, as we have already shown, equal to unity.

Draw horizontal lines through A and D to meet the line B C produced in K and Q, then, since the parallelograms A B C D and A K Q D stand on the same base and are between the same parallels, they are equal. Now draw the vertical lines A k and K P to meet Q D, produced if necessary. Then the rectangle A K P k is equal to the parallelogram A K Q D, because they stand on the same base A K, and are between the same parallels A K and k Q. Hence the rectangle A K P k

FIG. 24.

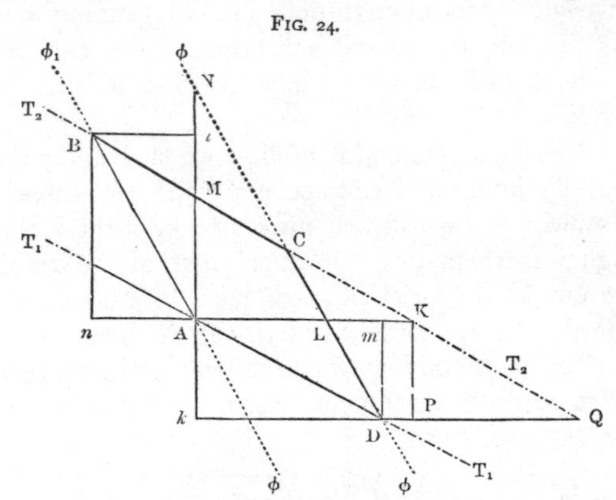

is also equal to the original parallelogram A B C D. If, therefore, we draw A K from A horizontally to meet the isothermal T_2, and A k vertically to meet a horizontal line through D, we shall have the following relation:

$$\text{A K . A } k = \text{A B C D.}$$

In the same way, if the horizontal line through A cuts the adiabatic line ϕ_2 in L and the verticals through D and B in m and n, and if the vertical line through A cuts the isothermal line T_2 in M, the adiabatic line ϕ_2 in N, and the horizontal line through B in l, we shall get the following four values of the area of A B C D, including that which we have already investigated:

$$\text{A B C D} = \text{A K . A } k = \text{A L . A } l = \text{A M . A } m = \text{A N . A } n = 1.$$

We have next to interpret the physical meaning of the four pairs of lines which enter into these products.

We must remember that the volume of the substance is measured horizontally to the right, and its pressure vertically upwards; that the interval between the isothermal lines represents one degree of temperature, the graduation of the scale being as much subdivided as we please; and that the interval between the adiabatic lines represents the addition of a quantity of heat whose numerical value is T, the absolute temperature.

(1) A K represents the increase of volume for a rise of temperature equal to one degree, the pressure being maintained constant. This is called the *dilatability* of the substance per unit of mass, and if we denote the dilatability per unit of volume by α, A K will be denoted by V α.

A k represents the diminution of pressure corresponding to the addition of a quantity of heat represented numerically by T, the temperature being maintained constant.

If the pressure is increased by unity, the temperature remaining constant, the quantity of heat which is emitted by the substance is $\frac{\mathrm{T}}{\mathrm{A}\,k}$. Since $\mathrm{A\,K} \,.\, \mathrm{A}\,k = 1$, $\frac{\mathrm{T}}{\mathrm{A}\,k} = \mathrm{T} \,.\, \mathrm{A\,K}$.

Hence the following relation between the dilatation under constant pressure and the heat developed by pressure.

First Thermodynamic Relation.—If the pressure of a substance be increased by unity while the temperature is maintained constant, the quantity of heat emitted by the substance is equal to the product of the absolute temperature into the dilatation for one degree of temperature under constant pressure.

Hence, if the temperature is maintained constant, those substances which increase in volume as the temperature rises give out heat when the pressure is increased, and those which contract as the temperature rises absorb heat when the pressure is increased.

(2) A L represents the increase of volume under constant pressure when a quantity of heat numerically equal to T is communicated to the substance.

A l represents the increase of pressure required to raise

the substance one degree of temperature when no heat is allowed to escape.

Second Thermodynamic Relation.—The quantity $\frac{\text{T}}{\text{A L}}$ represents the heat which must be communicated to the substance in order to increase its volume by unity, the pressure being constant. This is equal to the product of the absolute temperature into the increase of pressure required to raise the temperature one degree when no heat is allowed to escape.

(3) A M represents the increase of pressure corresponding to a rise of one degree of temperature, the volume being constant. (We may suppose the substance enclosed in a vessel the sides of which are perfectly unyielding.)

A *m* represents the increase of volume produced by the communication of a quantity of heat numerically equal to T, the temperature being maintained constant.

The heat given out by the substance when the volume is diminished by unity, the temperature being maintained constant, is therefore $\frac{\text{T}}{\text{A}\,m}$. This quantity is called the latent heat of expansion.

Since $\text{A M} \,.\, \text{A}\,m = 1$, we may express the relation between these lines thus: $\frac{\text{T}}{\text{A}\,m} = \text{T} \,.\, \text{A M}$, or, in words:

Third Thermodynamic Relation.—The latent heat of expansion is equal to the product of the absolute temperature and the increment of pressure per degree of temperature at constant volume.

(4) A N represents the increase of the pressure when a quantity, T, of heat is communicated to the substance, the volume being constant.

A *n* represents the diminution of volume when the substance, being prevented from losing heat, is compressed till the temperature rises one degree. Hence:

Fourth Thermodynamic Relation.—$\frac{1}{A\,n}$ represents the rise of temperature due to a diminution of the volume by unity, no heat being allowed to escape, and this is equal to A N, the increase of pressure at constant volume due to a quantity of heat, numerically equal to T, communicated to the substance.

We have thus obtained four relations among the physical properties of the substance. These four relations are not independent of each other, so as to rank as separate truths. Any one might be deduced from any other. The equality of the products A K, A k, &c., to the parallelogram A B C D and to each other is a merely geometrical truth, and does not depend upon thermodynamical principles. What we learn from thermodynamics is that the parallelogram and the four products are each equal to unity, whatever be the nature of the substance or its condition as to pressure and temperature.[1]

ON THE TWO MODES OF MEASURING SPECIFIC HEAT.

The quantity of heat required to raise unit of mass of the substance one degree of temperature is called the specific heat of the substance.

[1] These four relations may be concisely expressed in the language of the Differential Calculus as follows:

$$\frac{dv}{d\theta}(p \text{ const.}) = -\frac{d\phi}{dp}(\theta \text{ const.}) \quad . \quad . \quad . \quad (1)$$

$$\frac{dv}{d\phi}(p \text{ const.}) = \frac{d\theta}{dp}(\phi \text{ const.}) \quad . \quad . \quad . \quad (2)$$

$$\frac{dp}{d\theta}(v \text{ const.}) = \frac{d\phi}{dv}(\theta \text{ const.}) \quad . \quad . \quad . \quad (3)$$

$$\frac{dp}{d\phi}(v \text{ const.}) = -\frac{d\theta}{dv}(\phi \text{ const.}) \quad . \quad . \quad . \quad (4)$$

Here v denotes the volume.

p ,, pressure.

θ ,, absolute temperature.

ϕ ,, thermodynamic function.

If this quantity of heat is expressed in dynamical measure it is called the dynamical specific heat. It is more usually expressed in terms of the thermal unit as defined in the chapter on Calorimetry with reference to water. To reduce this to dynamical measure we must multiply by Joule's mechanical equivalent of the thermal unit.

But the specific heat of a substance depends on the mode in which the pressure and volume of the substance vary during the rise of temperature.

There are, therefore, an indefinite number of modes of defining the specific heat. Two only of these are of any practical importance. The first method is to suppose the volume to remain constant during the rise of temperature. The specific heat under this condition is called the specific heat at constant volume. We shall denote it by K_v.

In the diagram the line A M N represents the different states of the substance when the volume is constant, A M represents the increase of pressure due to a rise of one degree of temperature, and A N that due to the application of a quantity of heat numerically equal to T. Hence to find the quantity of heat, K_v, which must be communicated to the substance in order to raise its temperature one degree, and so increase the pressure by A M, we have

$$AN : AM :: T : K_v$$

or

$$K_v = T \cdot \frac{AM}{AN}.$$

The second method of defining specific heat is to suppose the pressure constant. The specific heat under constant pressure is denoted by K_p.

The line A L K in the diagram represents the different states of the substance at constant pressure, A K represents the increase of volume due to a rise of one degree of temperature, and A L represents the increase of volume due to a quantity of heat numerically equal to T. Now the quantity K_p of heat raises the substance one degree, and therefore increases the volume by A K.

Hence

$$\text{AL} : \text{AK} :: \text{T} : \text{K}_p$$

or

$$\text{K}_p = \text{T}\,\frac{\text{AK}}{\text{AL}}.$$

(A third mode of defining specific heat is sometimes adopted in the case of saturated steam. In this case the steam is supposed to remain at the point of saturation as the temperature rises. It appears, from the experiments of M. Regnault, as shown in the diagram at p. 135, that heat leaves the saturated steam as its temperature rises, so that its specific heat is *negative*, a result pointed out by Clausius and Rankine.)

ON THE TWO MODES OF MEASURING ELASTICITY.

The elasticity of a substance was defined at p. 107 to be the ratio of the increment of pressure to the compression produced by it, the compression being defined to be the ratio of the diminution of volume to the original volume.

But we require to know something about the thermal conditions under which the substance is placed before we can assign a definite value to the elasticity. The only two conditions which are of practical importance are, first, when the temperature remains constant, and, second, when there is no communication of heat.

(1) The elasticity under the condition that the temperature remains constant may be denoted by E_θ.

In this case the relation between volume and pressure is defined by the isothermal line D A. The increment of pressure is k A, and the diminution of volume is m A. Calling the volume V, the elasticity at constant temperature is

$$\text{E}_\theta = \text{V}\frac{\text{A}k}{\text{A}m} = \text{V}\,.\,\frac{\text{AM}}{\text{AK}}.$$

(2) The elasticity under the condition that heat neither enters nor leaves the substance is denoted by E_ϕ.

In this case the relation between volume and pressure is

defined by the adiabatic line A B. The increment of pressure is $A\,l$, and the decrement of volume is $A\,n$. Hence the elasticity when no heat escapes is

$$E_\phi = V\,.\,\frac{A\,l}{A\,n} = V\,.\,\frac{A\,N}{A\,L}.$$

There are several important relations among these quantities. In the first place, we find for the ratio of the specific heats,

$$\frac{K_p}{K_v} = \frac{T\,.\,\dfrac{A\,K}{A\,L}}{T\,.\,\dfrac{A\,M}{A\,N}} = \frac{V\,.\,\dfrac{A\,N}{A\,L}}{V\,.\,\dfrac{A\,M}{A\,K}} = \frac{E_\phi}{E_\theta}$$

or the ratio of the specific heat at constant pressure to that at constant volume is equal to the ratio of the elasticity when no heat escapes to the elasticity at constant temperature. This relation is quite independent of the principles of thermodynamics, being a direct consequence of the definitions.

The ratio of K_p to K_v, or of E_ϕ to E_θ, is commonly denoted by the symbol γ : thus $K_p = \gamma K_v$, and $E_\phi = \gamma E_\theta$.

Let us next determine the difference between the two elasticities

$$E_\phi - E_\theta = V\,.\,\frac{A\,l\,.\,A\,m - A\,n\,.\,A\,k}{A\,m\,.\,A\,n}.$$

The numerator of the fraction is evidently, by the geometry of the figure, equal to the parallelogram A B C D. Multiplying by K_v, we find

$$K_v\,(E_\phi - E_\theta) = T\,V\,.\,\frac{A\,M}{A\,m}\,.\,\frac{A\,B\,C\,D}{A\,N\,.\,A\,n} = T\,.\,V\,.\,\frac{A\,M}{A\,m},$$

since $A\,n\,.\,A\,N = A\,B\,C\,D$, as we have shown.

Since $K_v\,E_\phi = K_p\,E_\theta$, we also find

$$E_\theta\,(K_p - K_v) = T\,V\,.\,\frac{A\,M}{A\,m}.$$

These relations are independent of the principles of thermodynamics.

If we now apply the thermodynamical equation A M . A m = 1, each of these quantities becomes equal to

$$\text{T V} . (\text{A M})^2.$$

Now A M is the increment of pressure at constant volume per degree of temperature, a very important quantity. The results therefore may be written

$$K_v (E_\phi - E_\theta) = \text{T V A M}^2 = E_\theta (K_p - K_v).$$

CHAPTER X.

ON LATENT HEAT.

A VERY important class of cases is that in which the substance is in two different states at the same temperature and pressure, as when part of it is solid and part liquid, or part solid or liquid and part gaseous.

In such cases the volume occupied by the substance must be considered as consisting of two parts, v_1 being that of the substance in the first state, and v_2 that of the substance in the second state. The quantity of heat necessary to convert unit of mass of the substance from the first state to the second without altering its temperature is called the Latent Heat of the substance, and is denoted by L.

During this process the volume changes from v_1 to v_2 at the constant pressure p.

Let P S be an isothermal line, which in this case is horizontal, and let it correspond to the pressure P and the temperature S.

FIG. 25.

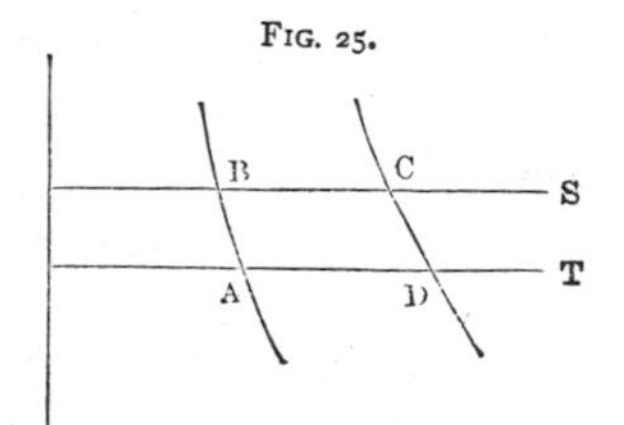

Let Q T be another isothermal line corresponding to the pressure Q and the temperature T.

Let B A and C D be adiabatic lines cutting the isothermals in A B C D.

Then the substance, in expanding at the temperature S from the volume P B to the volume P C, will absorb a quantity of heat equal to $L \frac{BC}{v_2 - v_1}$, where L is the latent heat at temperature S.

When the substance is compressed from Q D to Q A at temperature T it will give out a quantity of heat equal to

$$L' \frac{AD}{v_2' - v_1'},$$

where the accented quantities refer to the temperature T.

The quantity of work done by an engine when the indicating point describes the figure A B C D on the diagram is represented by the area of this figure, and if the temperatures S and T are so near each other that we may neglect the curvature of the lines A B and C D, this area is

$$\tfrac{1}{2}(BC + AD)\,PQ.$$

If the difference of pressures P Q is very small, B C = A D nearly, so that we may write the area thus:

$$BC\,(P - Q).$$

But we may calculate the work in another way. It is equal to the heat absorbed at the higher temperature, multiplied by the ratio of the difference of the temperatures to the higher temperature. This is

$$L \frac{BC}{v_2 - v_1} \quad \frac{S - T}{S}.$$

Equating the two values of the work, we find the latent heat

$$L = (v_2 - v_1)\,S\,\frac{P - Q}{S - T},$$

where it is to be remembered that in calculating the fraction $\frac{P - Q}{S - T}$ the difference of the pressures P and Q and the difference of the temperatures S and T are to be supposed very small. In fact, this fraction is that which in the language of the differential calculus would be denoted by $\frac{dp}{d\theta}$. The student may deduce the equation at once from the third thermodynamic relation at p. 160.

The most important case of a substance in two different states is that in which the substance is partly water and partly steam at the same temperature.

The pressure of steam in a vessel containing water at the same temperature is called the pressure of saturated steam or aqueous vapour at that temperature.

The value of this pressure has been determined for a great number of temperatures as measured on the ordinary scales. The most complete determinations of this kind are those of Regnault. Regnault has also determined L, the latent heat of unit of mass of steam, for many different temperatures.

Hence, if we also knew the value of $v_2 - v_1$, or the difference of volume between unit of mass of water and the same when converted into steam, we should have all the data for determining S, the absolute temperature on the thermodynamic scale.

Unfortunately there is considerable difficulty in ascertaining the volume of steam at the point of saturation. If we place a known weight of water in a vessel, the capacity of which we can adjust, and determine either the capacity corresponding to a given temperature at which the whole is just converted into steam, or the temperature corresponding to a given capacity, we may obtain data for determining the density of saturated steam, but it is exceedingly difficult to observe either the completion of the evaporation or the beginning of the condensation, and at the same time to avoid other causes of error. It is to be hoped that these difficulties will be overcome, and then our knowledge of the other properties of saturated steam will enable us to compare the ordinary scales of temperature with the thermodynamic scale through a range extending from $-30°$ F. to $432°$ F.

In the meantime Clausius and Rankine have made use of

the formula in order to calculate the density of saturated steam, assuming that the absolute temperature is equal to the temperature reckoned from $-460°$ of Fahrenheit's scale.

The same principle enables us to establish relations between the physical properties of a substance at the point at which it changes from the solid to the liquid state.

The temperature of melting ice was always supposed to be absolutely constant till it was pointed out by Professor James Thomson[1] that it follows from Carnot's principle that the melting point must be lowered when the pressure increases; for if v_1 is the volume of a pound of ice, and v_2 that of a pound of water, both being at 32° F., we know that the volume of the ice is greater than that of the water. Hence if S be the melting point at pressure P, and T the melting point at pressure Q, we have, as at p. 172,

$$\frac{S - T}{P - Q} = (v_2 - v_1)\frac{S}{L}.$$

If we make $P = 0$ and $S = 32°$ F., then the melting temperature at pressure Q is

$$T = 32° \text{ F.} - (v_1 - v_2)\, Q\, \frac{S}{L}.$$

Now the volume of a pound of ice at 32° F. is 0·0174 cubic feet $= v_1$, and that of a pound of water at the same temperature is 0·016 cubic feet $= v_2$. S, the absolute temperature, corresponding to 32° F., is 492°. L, the latent heat required to convert a pound of ice into a pound of water, = 142 thermal units = 142 × 772 foot-pounds. Hence T, the temperature of melting, corresponding to a pressure of Q pounds' weight per square foot, is

$$T = 32° - 0°{\cdot}0000063 \times Q.$$

If the pressure be that of n atmospheres, each atmosphere being 2,116 pounds' weight per square foot,

$$T = 32° - 0°{\cdot}0133\, n.$$

[1] *Transactions of the Royal Society of Edinburgh*, vol. xvi. p. 575, January 2, 1849.

Hence the melting point of ice is lowered by about the seventy-fifth part of a degree of Fahrenheit for every additional atmosphere of pressure. This result of theory was verified by the direct experiments of Professor W. Thomson.[1]

Professor J. Thomson has also pointed out the importance of the unique condition as to temperature and pressure under which water or any other substance can permanently exist in the solid, liquid, and gaseous forms in the same vessel. This can only be at the freezing temperature corresponding to the pressure of vapour at this freezing point. He calls this the triple point, because three thermal lines meet in it—(1) the steam line, which divides the liquid from the gaseous state; (2) the ice line, which divides the liquid from the solid state; (3) the hoar-frost line, which divides the solid from the gaseous state.

Whenever the volume of the substance is, like that of water, less in the liquid than in the solid state, the effect of pressure on a vessel containing the substance partly in a liquid and partly in a solid state is to cause some of the solid portion to melt, and to lower the temperature of the whole to the melting point corresponding to the pressure. If, on the contrary, the volume of the substance is greater in the liquid than in the solid state, the effect of pressure is to solidify some of the liquid part, and to raise the temperature to the melting point corresponding to the pressure. To determine at once whether the volume of the substance is greater in the liquid or the solid state, we have only to observe whether solid portions of the substance sink or swim in the melted substance. If, like ice in water, they swim, the volume is greater in the solid state, and pressure causes melting and lowers the melting point. If, like sulphur, wax, and most kinds of stone, the solid substance sinks in the liquid, then pressure causes solidification and raises the melting point.

[1] *Proceedings of the Royal Society of Edinburgh*, 1850.

When two pieces of ice at the melting point are pressed together, the pressure causes melting to take place at the portions of the surface in contact. The water so formed escapes out of the way and the temperature is lowered. Hence as soon as the pressure diminishes the two parts are frozen together with ice at a temperature below 32°. This phenomenon is called Regelation.

It is well known that the temperature of the earth increases as we descend, so that at the bottom of a deep boring it is considerably hotter than at the surface. We shall see that, unless we suppose the present state of things to be of no great antiquity, this increase of temperature must go on to much greater depths than any of our borings. It is easy on this supposition to calculate at what depth the temperature would be equal to that at which most kinds of stone melt in our furnaces, and it has been sometimes asserted that at this depth we should find everything in a state of fusion. But we must recollect that at such depths there is an enormous pressure, and therefore rocks which in our furnaces would be melted at a certain temperature may remain solid even at much greater temperatures in the heart of the earth.

CHAPTER XI.

ON THE APPLICATION OF THE PRINCIPLES OF THERMODYNAMICS TO GASES.

THE physical properties of bodies in the gaseous state are more simple than when they are in any other state. The relations of the volume, pressure, and temperature are then more or less accurately represented by the laws of Boyle and Charles, which we shall speak of, for brevity, as the 'gaseous laws.' We may express them in the following form:

Let v denote the volume of unit of mass, p the pressure, t the temperature measured by an air thermometer and reckoned from the absolute zero of that instrument, then the quantity $\frac{vp}{t}$ remains constant for the same gas.

We here use the symbol t to denote the absolute temperature as measured by the air thermometer, reserving the symbol θ to denote the temperature according to the absolute thermodynamic scale.

We have no right to assume without proof that these two quantities are the same, although we shall be able to show by experiment that the one is nearly equal to the other.

It is probable that when the volume and the temperature are sufficiently great all gases fulfil with great accuracy the gaseous laws; but when, by compression and cooling, the gas is brought near to its point of condensation into the liquid form, the quantity $\frac{vp}{t}$ becomes less than it is for the perfectly gaseous state, and the substance, though still apparently gaseous, no longer fulfils with accuracy the gaseous laws. (See pp. 116, 119.)

The specific heat of a gas can be determined only by a course of experiments involving considerable difficulty and requiring great delicacy in the measurements. The gas must be enclosed in a vessel, and the density of the gas itself is so small that its capacity for heat forms but a small part of the total capacity of the apparatus. Any error, therefore, in the determination of the capacity either of the vessel itself or of the vessel with the gas in it will produce a much larger error in the calculated specific heat of the gas.

Hence the determinations of the specific heat of gases were generally very inaccurate, till M. Regnault brought all the resources of his experimental skill to bear on the investigation, and, by making the gas pass in a continuous current and in large quantities through the tube of his calorimeter, deduced results which cannot be far from the truth.

These results, however, were not published till 1853, but in the meantime Rankine, by the application of the principles of thermodynamics to facts already known, determined theoretically a value of the specific heat of air, which he published in 1850. The value which he obtained differed from that which was then received as the best result of direct experiment, but when Regnault's result was published it agreed exactly with Rankine's calculation.

We must now explain the principle which Rankine applied. When a gas is compressed while the temperature remains constant, the product of the volume and pressure remains constant. Hence, as we have shown, the elasticity of the gas at constant temperature is numerically equal to its pressure.

But if the vessel in which the gas is contained is incapable of receiving heat from the gas, or of communicating heat to it, then when compression takes place the temperature will rise, and the pressure will be greater than it was in the former case. The elasticity, therefore, will be greater in the case of no thermal communication than in the case of constant temperature.

To determine the elasticity under these circumstances in this way would be impossible, because we cannot obtain a vessel which will not allow heat to escape from the gas within it. If, however, the compression is effected rapidly, there will be very little time for the heat to escape, but then there will be very little time to measure the pressure in the ordinary way. It is possible, however, after compressing air into a large vessel at a known temperature, to open an aperture of considerable size for a time which is sufficient to allow the air to rush out till the pressure is the same within and without the vessel, but not sufficient to allow much heat to be absorbed by the air from the sides of the vessel. When the aperture is closed the air is somewhat cooler than before, and though it receives heat from the sides of the vessel so fast that its temperature in the cooled state cannot be accurately observed with a thermometer, the amount of cooling may be calculated by observing the pressure of the air within the vessel after its temperature has become equal to that of the atmosphere. Since at the moment of closing the aperture the air within was cooler than the air without, while its pressure was the same, it follows that when the temperature within has risen so as to be equal to that of the atmosphere its pressure will be greater.

Let p_1 be the original pressure of the air compressed in a vessel whose volume is V; let its temperature be T, equal to that of the atmosphere.

Part of the air is then allowed to escape, till the pressure within the vessel is P, equal to that of the atmosphere; let the temperature of the air remaining within the vessel be t. Now let the aperture be closed, and let the temperature of the air within become again T, equal to that of the atmosphere, and let its pressure be then p_2.

To determine t, the absolute temperature of the air when cooled, we have, since the volume of the enclosed air is constant, the proportion

$$P : p_2 :: t : T,$$

or

$$t = \frac{P\,T}{p_2}.$$

This gives the cooling effect of expansion from the pressure p_1 to the pressure P. To determine the corresponding change of volume we must calculate the volume originally occupied by the air which remains in the vessel.

At the end of the experiment it occupies a volume V, at a pressure p_2 and a temperature T. At the beginning of the experiment its pressure was p_1 and its temperature T: hence the volume which it then occupied was $V\frac{p_2}{p_1} = v$, and a sudden increase of volume in the ratio of p_2 to p_1 corresponds to a diminution of pressure from p_1 to P. Since p_2 is greater than P, the ratio of the pressures is greater than the ratio of the volumes.

The elasticity of the air under the condition of no thermal communication is the value of the quantity

$$\frac{V + v}{2} \cdot \frac{p_1 - P}{V - v} \quad \text{or} \quad \tfrac{1}{2}(p_1 + p_2)\frac{p_1 - P}{p_1 - p_2}$$

when the expansion is very small, or when p_1 is very little greater than P.

But we know that the elasticity at constant temperature is numerically equal to the pressure (see p. 111). Hence we find for the value of γ, the ratio of the two elasticities,

$$\gamma = \frac{p_1 - P}{p_1 - p_2}.$$

or, more exactly,

$$\gamma = \frac{\log p_1 - \log P}{\log p_1 - \log p_2}.$$

Although this method of determining the elasticity in the case of no thermal communication is a practicable one, it is by no means the most perfect method. It is difficult, for instance, to arrange the experiment so that the pressure may be completely equalised at the time the aperture is closed, while at the same time no sensible portion of heat has been communicated to the air from the sides of the vessel. It is also necessary to ensure that no air has entered from without, and that the motion within the vessel has subsided before the aperture is closed.

But the velocity of sound in air depends, as we shall afterwards show, on the relation between its density and its pressure during the rapid condensations and rarefactions which occur during the propagation of sound. As these changes of pressure and density succeed one another several hundred, or even several thousand, times in a second, the heat developed by compression in one part of the air has no time to travel by conduction to parts cooled by expansion, even if air were as good a conductor of heat as copper is. But we know that air is really a very bad conductor of heat, so that in the propagation of sound we may be quite certain that the changes of volume take place without any appreciable communication of heat, and therefore the elasticity, as deduced from measurements of the velocity of sound, is that corresponding to the condition of no thermal communication.

The ratio of the elasticities of air, as deduced from experiments on the velocity of sound, is

$$\gamma = 1{\cdot}408.$$

This is also, as we have shown, the ratio of the specific heat at constant pressure to the specific heat at constant volume.

These relations were pointed out by Laplace, long before the recent development of thermodynamics.

We now proceed, following Rankine, to apply the thermodynamical equation of p. 171:

$$E_\theta (K_p - K_v) = T\,V\,(A\,M)^2.$$

In the case of a fluid fulfilling the gaseous laws, and also such that the absolute zero of its thermometric scale coincides with the absolute zero of the thermodynamic scale, we have

$$A\,M = \frac{p}{\theta}$$

and

$$E_\theta = p.$$

Hence

$$K_p - K_v = \frac{p\,v}{\theta} = R,$$

a constant quantity.

Now at the freezing temperature, which is $492^\circ{\cdot}6$ on Fahrenheit's scale from absolute zero, $p\,v = 26{,}214$

foot-pounds by Regnault's experiments on air, so that R is 53·21 foot-pounds per degree of Fahrenheit.

This is the work done by one pound of air in expanding under constant pressure while the temperature is raised one degree Fahrenheit.

Now K_v is the mechanical equivalent of the heat required to raise one pound of air one degree Fahrenheit without any change of volume, and K_p is the mechanical equivalent of the heat required to produce the same change of temperature when the gas expands under constant pressure, so that $K_p - K_v$ represents the additional heat required for the expansion. The equation, therefore, shows that this additional heat is mechanically equivalent to the work done by the air during its expansion. This, it must be remembered, is not a self-evident truth, because the air is in a different condition at the end of the operation from that in which it was at the beginning. It is a consequence of the fact, discovered experimentally by Joule (p. 196), that no change of temperature occurs when air expands without doing external work.

We have now obtained, in dynamical measure, the difference between the two specific heats of air.

We also know the ratio of K_p to K_v to be 1·408. Hence

$$K_v = \frac{53{\cdot}21}{{\cdot}408} = 130{\cdot}4 \text{ foot-pounds per degree Fahrenheit,}$$

and

$$K_p = K_v + 53{\cdot}21 = 183{\cdot}6 \text{ foot-pounds per degree Fah.}$$

Now the specific heat of water at its maximum density is Joule's equivalent of heat: for one pound it is 772 foot-pounds per degree Fahrenheit.

Hence if C_p is the specific heat of air at constant pressure referred to that of water as unity,

$$C_p = \frac{K_p}{J} = 0{\cdot}2378.$$

This calculation was published by Rankine in 1850.

The value of the specific heat of air, determined directly from experiment by M. Regnault and published in 1853, is

$$c_p = 0{\cdot}2379.$$

CHAPTER XII.

ON THE INTRINSIC ENERGY OF A SYSTEM OF BODIES.

THE energy of a body is its capacity for doing work, and is measured by the amount of work which it can be made to do. The Intrinsic energy of a body is the work which it can do in virtue of its actual condition, without any supply of energy from without.

Thus a body may do work by expanding and overcoming pressure, or it may give out heat, and this heat may be converted into work in whole or in part. If we possessed a perfect reversible engine, and a refrigerator at the absolute zero of temperature, we might convert the whole of the heat which escapes from the body into mechanical work. As we cannot obtain a refrigerator absolutely cold, it is impossible, even by means of perfect engines, to convert all the heat into mechanical work. We know, however, from Joule's experiments, the mechanical value of any quantity of heat, so that if we know the work done by expansion, and the quantity of heat given out by the body during any alteration of its condition, we can calculate the energy which has been expended by the body during the alteration.

As we cannot in any case deprive a body of all its heat, and as we cannot, in the case of bodies which assume the gaseous form, increase the volume of the containing vessel sufficiently to obtain all the mechanical energy of the expansive force, we cannot determine experimentally the whole energy of the body. It is sufficient, however, for all practical purposes to know how much the energy exceeds or falls short of the energy of the body in a certain definite condition—for instance, at a standard temperature and a standard pressure.

In all questions about the mutual action of bodies we are concerned with the difference between the energy of each body in different states, and not with its absolute value, so that the method of comparing the energy of the body at any time with its energy at the standard temperature and pressure is sufficient for our purpose. If the body in its actual state has less energy than when it is in the standard state, the expression for the relative energy will be negative. This, however, does not imply that the energy of a body can ever be really negative, for this is impossible. It only shows that in the standard state it has more energy than in the actual state.

Let us compare the energy of a substance in two different states. Let the two states be indicated in the diagram by the points A and B, and let the intermediate states through which it passes be indicated by the line, straight or curved, which is drawn from A to B.

During the passage from the state A to the state B the body does work by its pressure while expanding, and the quantity of energy expended in this way is measured by the area A B *b* *a* contained by the curve A B, the vertical lines from A and B, and the line of no pressure. If from A and B we draw adiabatic lines A α and B β, and produce them (in imagination) till they meet the line of absolute cold, then the heat given out by the substance during the change of state is represented by the area contained between the curve A B, the two adiabatic lines A α and B β, and the line of absolute cold, which probably coincides with the line of no pressure.

Fig. 26.

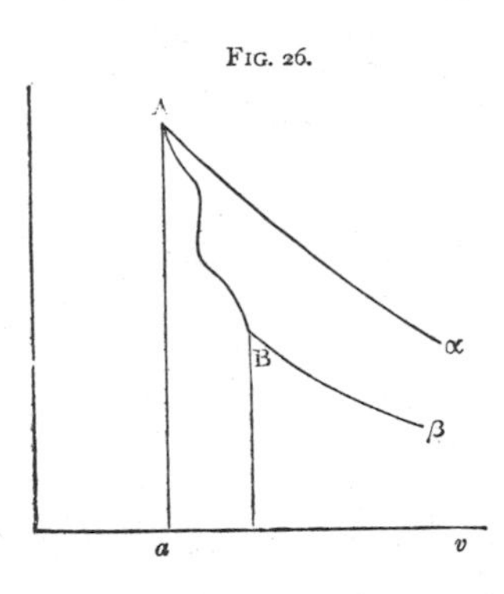

Now the energy of the substance in the state represented by A exceeds its energy in the state represented by B by the sum of the two portions of energy given out by the body in the form of mechanical work and in the form of heat, that is by a quantity represented by the sum of the two areas A B b a and A B β α. Now this is the quantity by which the area bounded by A a, A α, and the line of no pressure exceeds that bounded by B b, B β, and the line of no pressure. This therefore represents the excess of the energy in the state A over the energy in the state B.

Since we may suppose the point B on the line of no pressure, and as the energy would then be zero, we may define theoretically the total energy in the state represented by A as the area contained between the vertical line A a, the adiabatic line A α, and the line of no pressure a b v.

The excess of the intrinsic energy of the substance as thus defined when in a given state above its intrinsic energy at the standard temperature and pressure is the relative intrinsic energy of the substance, with which alone we have here to deal.

ON THE AVAILABLE ENERGY OF A SYSTEM OF BODIES.

Let us suppose that a number of different substances are placed in a confined region, the volume of which is V, and that no heat is allowed to escape from this region, though it may pass from one body to another within it.

Let us also suppose that we are able to make use of all the work done by the expansion of the different substances. Since they are in a confined space, one substance can expand only by compressing others, and work will be done only as long as the expanding substance has a greater pressure than those which it compresses. Hence, when all the substances are at the same pressure, no more work can be done in this way.

Finally, let us suppose that we are able by means of a perfect heat engine to transfer heat from one of the substances to another. Work will be done only when the heat is transferred from a hotter to a colder substance. Hence, when all the substances are reduced to the same temperature, no more work can be obtained in this way.

If, therefore, a number of substances are contained in a vessel which allows neither matter nor heat to pass its walls, the energy which can be converted into mechanical work will be entirely exhausted when all the substances are at the same pressure and the same temperature. To obtain any more work from the system, we must allow it to communicate either mechanically or thermally with bodies outside the vessel. Hence only a part of the whole intrinsic energy of the system is capable of being converted into mechanical work by actions going on within the vessel, and without any communication with external space by the passage either of matter or of heat. This part is sometimes called the Available Energy of the system. Clausius has called the remainder of the energy, which cannot be converted into work, the Entropy of the system. We shall find it more convenient to adopt the suggestion of Professor Tait, and give the name of Entropy to the part which can be converted into mechanical work.

DEFINITION OF ENTROPY.—*The Entropy of a system is the mechanical work it can perform without communication of heat, or alteration of its total volume, all transference of heat being performed by reversible engines.*

When the pressure and temperature of the system have become uniform the entropy is exhausted.

The original energy of the system is equal to the sum of the entropy and the energy remaining in the state of uniform pressure and temperature.

The entropy of a system consisting of several component systems is the same in whatever order the entropy of the parts is exhausted. It is therefore equal to the sum of the entropy of each component system, together with the entropy of the system consisting of the component systems, each with its own entropy exhausted.

When the parts of a system are at different temperatures, and if there is thermal communication between them, heat will pass from the hotter to the colder parts by conduction and radiation. The result of conduction and radiation is invariably to diminish the difference of temperature between the parts of the system, and the final effect is to reduce the whole system to a uniform temperature.

During this process no external mechanical work is done, and when the process is completed, and the temperature of the system has become uniform, no work can be obtained from the thermal energy of the system.

Hence the result of the conduction and radiation of heat from one part of a system to another is to diminish the entropy of the system, or the energy, available as work, which can be obtained from the system.

The energy of the system, however, is indestructible, and as it has not been removed from the system, it must remain in it. Hence the intrinsic energy of the system, when the entropy is exhausted by thermal communication, conduction, and radiation, is equal to its original energy, and is of course greater than in the case in which the entropy is exhausted by means of the reversible engine.

Again, when the parts of the system are at different pressures, and there is material communication by open channels between them, there will be a tendency of the parts of the system to move, and the result of this motion will be to equalise the pressure in the different parts. If the energy developed in this way is not gathered up and used in working a machine it will be spent in giving velocity to the parts of the system. But as soon as this motion is set up it begins to decay on account of the resistance which all substances, even in the gaseous state, offer to the relative motion of their parts. This resistance in the case of solid bodies sliding over each other is called Friction. In fluids it is called Internal Friction or Viscosity. In every case it tends to destroy the relative motion of the parts, and to convert the energy of this motion into heat.

If, therefore, the system contains portions of matter in which the pressure is different, and if there is a material communication (that is, an open passage of any kind) between these portions, the part of the entropy depending on difference of pressures will be converted first into visible motion of the parts, and, as this decays, into heat.

It is possible to prevent material communication between the parts of a system by enclosing the substances in vessels through which they cannot pass. But it is impossible to prevent thermal communication between the parts of the system, because no substance known to us is a non-conductor of heat. Hence the entropy of every system is in a state of decay unless it is supplied from without.

This is Thomson's doctrine of the Dissipation of Energy.[1] Energy is said to be dissipated when it cannot be rendered available as mechanical energy. The energy which is not yet dissipated is what we have here, following Tait, called Entropy. The theory of entropy, in this sense, was given by Thomson in 1853.[2] The name, however, was first employed by Clausius in 1854,[3] and it was used by him to denote the energy already dissipated.

The law of communication of heat, on which we founded our first definition of temperature; the principle of Carnot, and the second law of thermodynamics; and the theory of Dissipation of Energy, may be considered as expressions of the same natural fact with increasing degrees of scientific completeness. We shall return to the subject when we come to molecular theories.

[1] 'On a Universal Tendency in Nature to Dissipation of Energy.' —*Phil. Mag. and Proc. R.S.E.* 1852.

[2] 'On the Restoration of Mechanical Energy from an unequally heated Space.'—*Phil. Mag.* Feb. 1853.

[3] *Pogg. Ann.* Dec. 1854.

CHAPTER XIII.

ON FREE EXPANSION.

Theory of a Fluid in which no External Work is done during a Change of Pressure.

LET a fluid be forced through a small hole, or one or more narrow tubes, or a porous plug, and let the work done by the pressure from behind be entirely employed in overcoming the resistance of the fluid, so that when the fluid, after passing through the plug, has arrived at a certain point its velocity is very small. Let us also suppose that no heat enters or leaves the fluid, and that no sound or other vibration, the energy of which is comparable with that which would sensibly alter the temperature of the fluid, escapes from the apparatus.

We also suppose that the motion is steady—that is, that the same quantity of the fluid enters and issues from the apparatus in every second.

During the passage of unit of mass through the apparatus, if P and V are its pressure and volume at the section A before reaching the plug, and p, v the same at the section B after passing through it, the work done in forcing the fluid through the section A is P V, and the work done by the fluid in issuing through the section B is $p\ v$, so that the amount of work communicated to the fluid in passing through the plug is $\text{P V} - p\ v$.

FIG. 27.

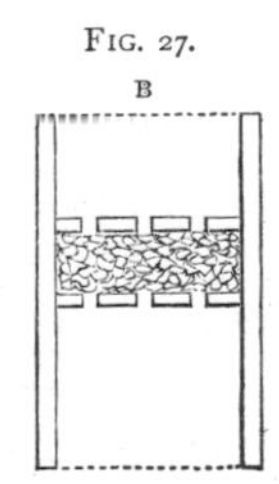

Hence, if E is the energy of unit of mass of the fluid while entering at the section A, and e the energy of unit of mass issuing at the section B,

$$e - \text{E} = \text{PV} - p v,$$

$$\text{E} + \text{P V} = e + p\ v \quad . \quad . \quad . \qquad (1)$$

That is to say, the sum of the intrinsic energy and the product of the volume and the pressure remains the same after passing through the plug, provided no heat is lost or gained from external sources.

Now the intrinsic energy E is indicated on the diagram by the area between A α an adiabatic line, A a a vertical line, and a b v the line of no pressure, and P V is represented by the rectangle A p O a. Hence the area included by α A p O v, the lines A α and O v being produced till they meet, represents the quantity which remains the same after passing through the plug. Hence in the figure the area A p q R is equal to the area contained between B R and the two adiabatic lines R α and B β.

FIG. 28.

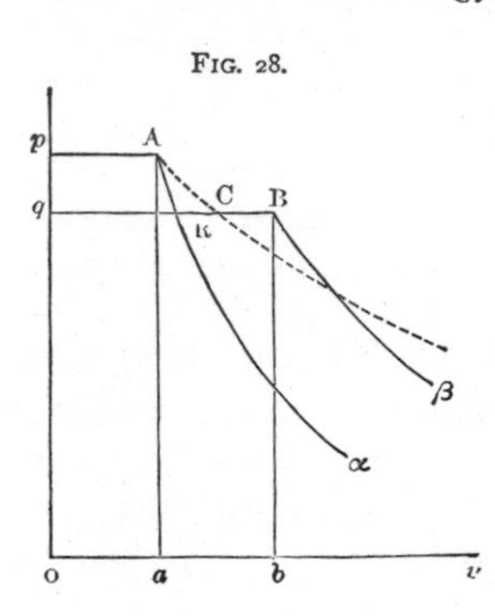

We shall next examine the relations between the different properties of the substance, in order to determine the rise of temperature corresponding to a passage through the plug from a pressure P to a pressure p, and we shall first suppose that P is not much greater than p.

Let A C be an isothermal line through A, cutting q B in C, and let us suppose that the passage of the substance from the state represented by A to the state represented by B is effected by a passage along the isothermal line A C, followed by an increase of volume from C to B. The smaller the distance A B, the less will the results of this process differ from those of the actual passage from A to B, in whatever manner this is really effected.

In passing from A to C, at the constant temperature θ, the pressure diminishes from P to p. The heat absorbed during this process is, by the first thermodynamic relation (p. 165),

$$(\text{P} - p)\ \text{V}\ \theta\ \alpha,$$

where α is the dilatation of unit of volume at constant pressure per degree of temperature.

In passing from C to B the substance expands at constant pressure, and its temperature rises from θ to $\theta + \tau$.

The heat required to produce this rise of temperature is

$$K_p \tau,$$

where K_p denotes the specific heat of the substance at constant pressure.

The whole heat absorbed by the substance during the passage from A B is therefore

$$(P - p) V \theta \alpha + K_p \tau,$$

and this is the value of the area between A B and the two adiabatic lines A α, B β.

Now this is equal to the area A p q B or $(P - p)$ V.

Hence we have the equation

$$K_p \tau = (P - p) V (1 - \theta \alpha) \quad . \quad . \quad . \quad (2)$$

where K_p denotes the specific heat of unit of mass at constant pressure, expressed in dynamical measure;

τ, the rise of temperature after passing through the plug;

$P - p$, the *small* difference of pressure on the two sides of the plug;

V, the volume of unit of mass (when $P - p$ is so great as to cause considerable alteration of volume, this quantity must be treated differently);

θ, the temperature on the absolute dynamical scale;

α, the dilatation of unit of volume at constant pressure per degree of temperature.

There are two cases in which observations of the rise (or fall) of temperature may be applied to determine quantities of great importance in the science of heat.

1. *To Determine the Dynamical Equivalent of Heat.*—The first case is that in which the substance is a liquid such as water or mercury, the volume of which is but slightly affected either by pressure or by temperature. In this case V will

vary so little that the effect of its variation may be taken into account as a correction required only in calculations of great accuracy. The dilatation a is also very small, so much so that the product θa, though not to be absolutely neglected, may be found with sufficient accuracy without a very accurate knowledge of the absolute value of θ.

If we suppose the pressure to be due to a depth of fluid equal to H on one side of the plug and h on the other, then

$$(P - p) = (H - h) \rho g,$$

where ρ is the density, and g is the numerical measure of the force of gravity. Now

$$V \rho = 1,$$

so that equation (2) becomes

$$K_p \tau = g (H - h) (1 - \theta a),$$

an equation from which we can determine K_p when we know τ the rise of temperature, and $H - h$ the difference of level of the liquid, a its coefficient of dilatation by heat, and (within a moderate degree of exactness) θ the absolute temperature in terms of the degrees of the same thermometer which is used to determine τ.

The quantity K_p is the specific heat at constant pressure, that is the quantity of heat which will raise unit of mass of the substance one degree of the thermometer. It is expressed here in absolute dynamical measure.

If the specific heat is to be expressed in gravitation measure, as in foot-pounds, we must divide by g, the force of gravity. If the specific heat is to be expressed in terms of the specific heat of a standard substance, as, for instance, water at its maximum density, we must divide by J, the specific heat of this substance.

We have already shown how by a direct experiment to compare the specific heat of any substance with that of water. If the specific heat expressed in this way is denoted by C_p, while K_p is the same quantity expressed in absolute measure, then the mechanical equivalent of the thermal unit is

$$J = \frac{K_p}{C_p}.$$

The quantity J is called Joule's Mechanical Equivalent of Heat, because Joule was the first to determine its value by an accurate method. It may be defined as the specific heat, in dynamical measure, of water at its maximum density.

It is equal to 772 foot-pounds at Manchester per pound of water. If we alter the standard of mass, we at the same time alter the unit of work in the same proportion, so that we must still express J by the same number. Hence we may express Joule's result by saying that the work done by any quantity of water in falling 772 feet at Manchester is capable of raising that water one degree Fahrenheit. If we wish to render the definition independent of the value of gravity at a particular place, we have only to calculate the velocity of a body after falling 772 feet at Manchester. The energy corresponding to this velocity in any mass of water is capable when converted into heat of raising the water one degree Fahrenheit.

There are considerable difficulties in obtaining the value of J by this method, even with mercury, for which a pressure of 25 feet gives a rise of one degree Fahrenheit.

2. *To Determine the Absolute Zero of the Thermodynamic Scale of Temperature.*—The most important application of the method is to determine the absolute zero of temperature on the thermodynamic scale as compared with the absolute zero of an air thermometer.

For this purpose, air or any other gas which fulfils with sufficient accuracy the laws of Boyle and Charles is forced in a steady current through a porous plug, and the difference of temperatures on the two sides of the plug is observed.

In this case we have $\frac{v\,p}{t}$ constant, and $\alpha = \frac{1}{t}$, so that

$$K_p \tau = \frac{v_0 p_0}{t_0} \frac{P - p}{p} (t - \theta).$$

In this expression t is the absolute temperature indicated by a thermometer consisting of the gas employed in the experiment, and θ is the absolute temperature on the dynamical scale.

The expression is true only when $P - p$ is small compared with P or p. When, as is the case in actual experiments, P is several times p, we must consider the pressure as sinking from P to p by a series of small steps, and trace the change of temperature from step to step. To do this requires a process of the same nature as that described in our calculation of the height of a mountain (Chapter XIV.) The result is that, instead of $\frac{P - p}{p}$, we must write $\log_\varepsilon \frac{P}{p}$, where the logarithm is Napierian, or

$$2\cdot30258 \log \frac{P}{p}$$

where the logarithm is that of the common tables;

whence
$$\theta = t - \frac{t}{v p} \cdot \frac{K_p \tau}{2\cdot3026 \log \frac{P}{p}}$$

an expression which gives the absolute temperature θ in terms of the temperature t reckoned from the absolute zero of a thermometer constructed of the gas used in the experiment. Here K_p is the specific heat of the gas at constant pressure, and τ is the increase of temperature after passing through the porous plug.

In the case of most of the gases examined by Joule and Thomson there was a slight cooling effect on the gas passing through the plug. In other words, τ was negative, and the absolute temperature was therefore higher than that indicated by the gaseous thermometer. The ratio, therefore, in which the gas expanded between two standard temperatures was greater than the true ratio of these temperatures on the thermodynamic scale. The cooling effect was much greater with carbonic acid than with oxygen, nitrogen, or air, as was to be expected, because we know from the experiments of Regnault that the dilatation of carbonic acid is greater than that of air or its constituents. It was also found, for all these gases, that the cooling effect was less at high temperatures, which shows that as the temperature rises the dilatation of the gas is more and more accurately proportional to the absolute temperature of the thermodynamic scale.

The only gas which exhibited a contrary effect was hydrogen, in which there was a slight heating effect after passing the plug.

The result of the experiments of Joule and Thomson was to show that the temperature of melting ice is 273°·7 on the thermodynamic scale, the degrees being such that there are 100 of them between this temperature and that of the vapour of boiling water at the standard pressure.

The absolute zero of the thermodynamic scale is therefore −273°·7 Centigrade, or −460°·66 Fahrenheit.

It appears, therefore, that, in the more perfect gases, the cooling effect due to expansion is almost exactly balanced by the heating effect due to the work done by the expansion when this work is wholly spent in generating heat in the gas. This result had been already obtained, although by a method not admitting of such great accuracy, by Joule,[1] who showed that the intrinsic energy of a gas is the same at the same temperature, whatever be the volume which it occupies.

To test this, he compressed air into a vessel till it contained about 22 atmospheres, and exhausted the air from another vessel. These vessels were then connected by

[1] *Phil. Mag.* May 1845.

means of a pipe closed by a stopcock, and the whole placed in a vessel of water.

After a sufficient time the water was thoroughly stirred, and its temperature taken by means of a delicate thermometer. The stopcock was then opened by means of a proper key, and the air allowed to pass from the full into the empty vessel till equilibrium was established between the two. Lastly the water was again stirred and its temperature carefully noted.

From a number of experiments of this kind, carefully corrected for all sources of error, Joule was led to the conclusion that *no change of temperature occurs when air is allowed to expand in such a manner as not to develop mechanical power.*

This result, as has been shown by the more accurate experiments afterwards made by Joule and W. Thomson, is not quite correct, for there is a slight cooling effect. This effect, however, is very small in the case of permanent gases, and diminishes when the gas, by rise of temperature or diminution of pressure, approaches nearer to the condition of a perfect gas.

We may however assert, as the result of these experiments, that the amount of heat absorbed by a gas expanding at uniform temperature is nearly, though not exactly, the thermal equivalent of the mechanical work done by the gas during the expansion. In fact, we know that in the case of air the heat absorbed is a little greater and in hydrogen a very little less than this quantity.

This is a very important property of gases. If we reverse the process, we find that the heat developed by compressing air at constant temperature is the thermal equivalent of the work done in compressing it.

This is by no means a self-evident proposition. In fact, it is not true in the case of substances which are not in the gaseous state, and even in the case of the more imperfect gases it deviates from the truth. Hence the calculation of the dynamical equivalent of heat, which Mayer founded on this proposition, at a time when its truth had not been experimentally proved, cannot be regarded as legitimate.

CHAPTER XIV.

ON THE DETERMINATION OF HEIGHTS BY THE BAROMETER.

THE barometer is an instrument by means of which the pressure of the air at a particular place may be measured. In the mercurial barometer, which is the most perfect form of the instrument, the pressure of the air on the free surface of the mercury in the cistern is equal to that of a column of mercury whose height is the difference between the level of the mercury in the cistern, which sustains the pressure of the air, and that of the mercury in the tube, which has no air above it. The pressure of the air is often expressed in terms of the height of this column. Thus we speak of a pressure of 30 inches of mercury, or of a pressure of 760 millimetres of mercury.

To express a pressure in absolute measure we must consider the force exerted against unit of area. For this purpose we must find the weight of a column of mercury of the given height standing on unit of area as base.

If h is the height of the column, then, since its section is unity, its volume is expressed by h.

To find the mass of mercury contained in this volume we must multiply the volume by the density of mercury. If this density is denoted by ρ, the mass of the column is ρh. The pressure, which we have to find, is the force with which this mass is drawn downwards by the earth's attraction. If g denotes the force of the earth's attraction on unit of mass, then the force on the column will be $g \rho h$. The pressure therefore of a column of mercury of height h is expressed by

$$g \rho h,$$

where h is the height of the column, ρ the density of mercury, and g the intensity of gravity at the place. The density of mercury diminishes as the temperature increases. It is usual to reduce all pressures measured in this way to the height of a column of mercury at the freezing temperature of water.

If two barometers at the same place are kept at different temperatures, the heights of the barometers are in the proportion of the volumes of mercury at the two temperatures.

The intensity of gravitation varies at different places, being less at the equator than at the poles, and less at the top of a mountain than at the level of the sea.

It is usual to reduce observed barometric heights to the height of a column of mercury at the freezing point and at the level of the sea in latitude 45°, which would produce the same pressure.

If there were no tides or winds, and if the sea and the air were perfectly calm in the whole region between two places, then the actual pressure of the air at the level of the sea must be the same in these two places; for the surface of the sea is everywhere perpendicular to the force of gravity. If, therefore, the pressure on its surface were different in two places, water would flow from the place of greater pressure to the place of less pressure till equilibrium ensued.

Hence, if in calm weather the barometer is found to stand at a different height in two different places at the level of the sea, the reason must be that gravity is more intense at the place where the barometer is low.

Let us next consider the method of finding the depth below the level of the sea by means of a barometer carried down in a diving bell.

If D is the depth of the surface of the water in the diving bell below the surface of the sea, and if p is the pressure of the atmosphere on the surface of the sea, then the pressure of the air in the diving bell must exceed that on the surface of the sea by the pressure due to a column of water of depth D. If σ is the density of sea-water, the pressure due to a column of depth D is $g \sigma D$.

Let the height of the barometer at the surface of the sea be observed, and let us suppose that in the diving bell it is found to be higher by a height h, then the additional pressure indicated by this rise is $g \rho h$, where ρ is the density of mercury. Hence

$$g \sigma D = g \rho h,$$

or

$$D = \frac{\rho}{\sigma} h = s h,$$

where $s = \frac{\rho}{\sigma} = \frac{\text{density of mercury}}{\text{density of water}}$ = specific gravity of mercury.

The depth below the surface of the sea is therefore equal to the product of the rise of the barometer multiplied by the specific gravity of mercury. If the water is salt we must divide this result by the specific gravity of the salt water at the place of observation.

The calculation of depths under water by this method is comparatively easy, because the density of the water is not very different at different depths. It is only at great depths that the compression of the water would sensibly affect the result.

If the density of air had been as uniform as that of water, the measurement of heights in the atmosphere would have been as easy. For instance, if the density of air had been equal to σ at all pressures, then, neglecting the variation of gravity with height above the earth, we should find the height $\mathfrak{H}$ of the atmosphere thus: Let h be the height of the barometer, and ρ the density of mercury, then the pressure indicated by the barometer is

$$p = g \rho h.$$

If $\mathfrak{H}$ is the height of an atmosphere of density σ, it produces a pressure

$$p = g \sigma \mathfrak{H}.$$

Hence

$$\mathfrak{H} = h \frac{\rho}{\sigma}.$$

This is the height of the atmosphere above the place on the *false* supposition that its density is the same at all heights as it is at that place. This height is generally referred to as the *height of the atmosphere supposed of uniform density*, or more briefly and technically as the height of the *homogeneous atmosphere.*

Let us for a moment consider what this height (which evidently has nothing to do with the real height of the atmosphere) really represents. From the equation

$$p = g \sigma \mathfrak{H},$$

remembering that σ the density of air is the same thing as the reciprocal of v the volume of unit of mass, we get

$$\mathfrak{H} = \frac{p v}{g},$$

or $\mathfrak{H}$ is simply the product $p v$ expressed in gravitation measure instead of absolute measure.

Now, by Boyle's law the product of the pressure and the volume at a constant temperature is constant, and by Charles's law this product is proportional to the absolute temperature. For dry air at the temperature of melting ice, and when $g = 32 \cdot 2$,

$$\mathfrak{H} = \frac{p v}{g} = 26{,}214 \text{ feet},$$

or somewhat less than five statute miles.

It is well known that Mr. Glaisher has ascended in a balloon to the height of seven miles. This balloon was supported by the air, and though the air at this great height was more than three times rarer than at the earth's surface, it was possible to breathe in it. Hence it is certain that the atmosphere must extend above the height $\mathfrak{H}$, which we have deduced from our false assumption that the density is uniform.

But though the density of the atmosphere is by no means uniform through great ranges of height, yet if we confine ourselves to a very small range, say the millionth part of $\mathfrak{H}$—that is, about 0·026 feet, or less than the third of an inch—the density will only vary one-millionth part of itself from the top to the bottom of this range, so that we may suppose the pressure at the bottom to exceed that at the top by exactly one-millionth.

Let us now apply this method to determine the height of a mountain by the following imaginary process, too laborious to be recommended, except for the purpose of explaining the practical method :

We shall suppose that we begin at the top of the mountain, and that, besides our barometer, we have one thermometer to determine the temperature of the mercury, and another to determine the temperature of the air. We are also provided with a hygrometer, to determine the quantity of aqueous vapour in the air, so that by the thermometer and hygrometer we can calculate $\mathfrak{H}$, the height of the homogeneous atmosphere, at every station of our path.

On the top of the mountain, then, we observe the height of the barometer to be p. We now descend the mountain till we observe the mercury in the barometer to rise by one-millionth part of its own height. The height of the barometer at this first station is

$$p_1 = (1{\cdot}000001)\,p.$$

The distance we have descended is one-millionth of $\mathfrak{H}$, the height of the homogeneous atmosphere for the observed temperature at the first stage of the descent. Since it is at present impossible to measure pressures, &c., to one-millionth of their value, it does not matter whether $\mathfrak{H}$ be measured at the top of the mountain or one-third of an inch lower down.

Now let us descend another stage, till the pressure again increases one-millionth of itself, so that if p_2 is the new pressure,

$$p_2 = (1{\cdot}000001)\, p_1,$$

and the second descent is through a height equal to the millionth of $\mathfrak{H}_2$, the height of the homogeneous atmosphere in the second stage.

If we go on in this way n times, till we at last reach the bottom of the mountain, and if p_n is the pressure at the bottom,

$$\begin{aligned} p_n &= (1{\cdot}000001)\, p_{n-1} \\ &= (1{\cdot}000001)^2\, p_{n-2} \\ &= (1{\cdot}000001)^n\, p, \end{aligned}$$

and the whole vertical height will be

$$h = \frac{\mathfrak{H}_1 + \mathfrak{H}_2 + \&c. + \mathfrak{H}_n}{1{,}000{,}000}.$$

If we assume that the temperature and humidity are the same at all heights between the top and the bottom, then $\mathfrak{H}_1 = \mathfrak{H}_2 = \&c. = \mathfrak{H}_n = \mathfrak{H}$, and the height of the mountain will be

$$= \frac{n}{1{,}000{,}000}\, \mathfrak{H}.$$

If we know n, the number of stages, we can determine the height of the mountain in this way. But it is easy to find n without going through the laborious process of descending by distances of the third of an inch, for since $p_n = P$ is the pressure at the bottom, and p that at the top, we have the equation

$$P = (1{\cdot}000001)^n\, p.$$

Taking the logarithm of both sides of this equation, we get

$$\log P = n \log (1{\cdot}000001) + \log p,$$

or

$$n = \frac{\log P - \log p}{\log (1{\cdot}000001)}.$$

Now $\log 1{\cdot}000001 = 0{\cdot}0000004342942648$.

Substituting this value in the expression for h, we get

$$h = \frac{\mathfrak{H}}{{\cdot}434294} \log \frac{P}{p},$$

where the logarithms are the common logarithms to base 10, or

$$h = 2{\cdot}302585\, \mathfrak{H} \log \frac{P}{p}.$$

For dry air at the temperature of melting ice $\mathfrak{H} = 26{,}214$ feet : hence

$$h = \log \frac{P}{p} \times \left\{ 60360 + (\theta - 32^\circ)(122{\cdot}68) \right\}$$

gives the height in feet for a temperature θ on Fahrenheit's scale.

For rough purposes, the difference of the logarithms of the heights of the barometer multiplied by 10,000 gives the difference of the heights in fathoms of six feet.

CHAPTER XV.

ON THE PROPAGATION OF WAVES

THE following method of investigating the conditions of the propagation of waves is due to Prof. Rankine.[1] It involves only elementary principles and operations, but leads to results which have been hitherto obtained only by operations involving the higher branches of mathematics.

[1] *Phil. Trans.* 1869: 'On the Thermodynamic Theory of Waves of Finite Longitudinal Disturbance.'

The kind of waves to which the investigation applies are those in which the motion of the parts of the substance is along straight lines parallel to the direction in which the wave is propagated, and the wave is defined to be one which is propagated with constant velocity, and the type of which does not alter during its propagation.

In other words, if we observe what goes on in the substance at a given place when the wave passes that place, and if we suddenly transport ourselves a certain distance forward in the direction of propagation of the wave, then after a certain time we shall observe exactly the same things occurring in the same order in the new place, when the wave reaches it. If we travel with the velocity of the wave, we shall therefore observe no change in the appearance presented by the wave as it travels along with us. This is the characteristic of a wave of permanent type.

We shall first consider the quantity of the substance which passes in unit of time through unit of area of a plane which we shall suppose fixed, and perpendicular to the direction of motion.

Let u be the velocity of the substance, which we shall suppose to be uniform, then in unit of time a portion of the substance whose length is u passes through any section of a plane perpendicular to the direction of motion. Hence the volume which passes through unit of area is represented by u.

Now let Q be the quantity of the substance which passes through, and let v be the volume of unit of mass of the substance, then the whole volume is $Q\,v$, and this, by what we have said, is equal to u, the velocity of the substance. If the plane, instead of being fixed, is moving forwards with a velocity U, the quantity which passes through it will depend, not on the absolute velocity, u, of the substance, but on the relative velocity, $u - U$, and if Q is the quantity which passes through the plane from *right* to *left*,

$$Q\,v = U - u \qquad (1)$$

Let A be an imaginary plane moving from *left* to *right* with velocity U, and let this be the velocity of propagation

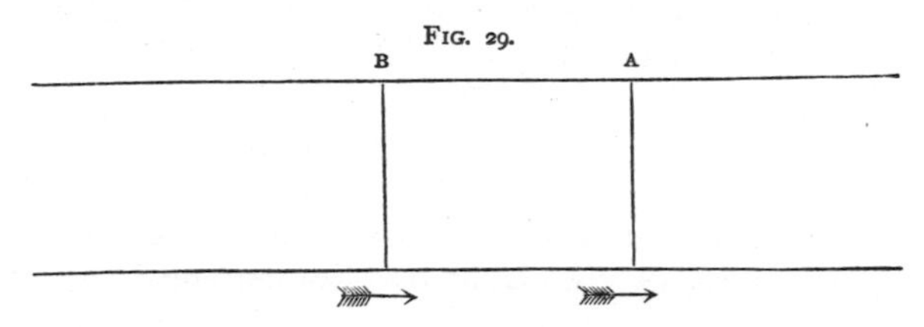

FIG. 29.

of the wave, then, as the plane A travels along, the values of u and all other quantities belonging to the wave at the plane A remain the same. If u_1 is the absolute velocity of the substance at A, v_1 the volume of unit of mass, and p_1 the pressure, all these quantities will be constant, and

$$Q_1 v_1 = U - u_1 \quad \ldots \ldots \ldots \quad (2)$$

If B be another plane, travelling with the same velocity U, and if Q_2 u_2 v_2 p_2 be the corresponding values at B,

$$Q_2 v_2 = U - u_2 \quad \ldots \ldots \ldots \quad (3)$$

The distance between the planes A and B remains invariable, because they travel with the same velocity. Also the quantity of the substance intercepted between them remains the same, because the density of the substance at corresponding parts of the wave remains the same as the wave travels along. Hence the quantity of matter which enters the space between A and B at A must be equal to that which leaves it at B, or

$$Q_1 = Q_2 = Q \text{ (say)} \quad \ldots \ldots \ldots \quad (4)$$

Hence

$$u_1 = U - Q v_1 \quad u_2 = U - Q v_2 \quad \ldots \quad (5)$$

so that when we know U and Q and the volume of unit of mass, we can find u_1 and u_2.

Let us next consider the forces acting on the matter contained between A and B. If p_1 is the pressure at A, and p_2

that at B, the force arising from these pressures tending to increase the momentum from left to right is $p_2 - p_1$.

This is the momentum generated in unit of time by the external pressures on the portion of the substance between A and B.

Now we must recollect that, though corresponding points of the substance in this interval are always moving in the same way, the matter itself between A and B is continually changing, a quantity Q entering at A, and an equal quantity Q leaving at B.

Now the portion Q which enters at A has a velocity u_1, and therefore a momentum Q u_1, and that which issues at B has a velocity u_2, and therefore a momentum Q u_2.

Hence the momentum of the entering fluid exceeds that of the issuing fluid by

$$Q(u_1 - u_2).$$

The only way in which this momentum can be produced is by the action of the external pressures p_1 and p_2; for the mutual actions of the parts of the substance cannot alter the momentum of the whole. Hence we find

$$p_2 - p_1 = Q(u_1 - u_2) \quad . \quad . \quad . \quad . \quad . \quad . \quad . \quad (6)$$

Substituting the values of u_1 and u_2 from equation (5), we find

$$p_2 - p_1 = Q^2(v_1 - v_2) \quad . \quad . \quad . \quad . \quad . \quad . \quad (7)$$

Hence

$$p_1 + Q^2 v_1 = p_2 + Q^2 v_2 \quad . \quad . \quad . \quad . \quad . \quad (8)$$

Now the only restriction on the position of the plane B is that it must remain at a constant distance behind A, and whatever be the distance between A and B, the above equation is always true.

Hence the quantity $p + Q^2 v$ must continue constant during the whole process involved in the passage of the wave. Calling this quantity P, we have

$$p = P - Q^2 v \quad . \quad . \quad . \quad . \quad . \quad . \quad . \quad . \quad . \quad . \quad (9)$$

or the pressure is equal to a constant pressure, P, diminished by a quantity proportional to the volume v.

This relation between pressure and volume is not fulfilled in the case of any actual substance. In all substances it is true that as the volume diminishes the pressure increases, but the increase of pressure is never strictly proportional to the diminution of volume. As soon as the diminution of volume becomes considerable, the pressure begins to increase in a greater ratio than the volume diminishes.

But if we consider only small changes of volume and pressure, we may make use of our former definition of elasticity at p. 107—namely, the ratio of the increase of pressure to the diminution of volume when the volume is unity, or, calling the elasticity E,

$$E = v \frac{p_2 - p_1}{v_1 - v_2} = v\,Q^2 \text{ by equation (7)} \quad (10)$$

where v is the volume of unit of mass, and since v_1 and v_2 are very nearly equal, we may take either for the value of v. Again, if v is the volume of unit of mass in those parts of the substance which are not disturbed by the wave, and for which, therefore, $u = 0$,

$$U = Q\,v \quad . \quad . \quad . \quad . \quad . \quad . \quad . \quad . \quad . \quad . \quad (11)$$

Hence we find

$$U^2 = Q^2\,v^2 = E\,v \quad . \quad . \quad . \quad . \quad . \quad . \quad . \quad . \quad (12)$$

which shows that the square of the velocity of propagation of a wave of longitudinal displacement in any substance is equal to the product of the elasticity and the volume of unit of mass.

In calculating the elasticity we must take into account the conditions under which the compression of the substance actually takes place. If, as in the case of sound-waves, it is very sudden, so that any heat which is developed cannot be conducted away, then we must calculate the elasticity on the supposition that no heat is allowed to escape.

In the case of air or any other gas the elasticity at constant temperature is numerically equal to the pressure. If we denote, as usual, the ratio of the specific heat at constant pressure to that at constant volume by the symbol γ, the elasticity when no heat escapes is

$$E_\phi = \gamma p \quad . \quad . \quad . \quad . \quad . \quad . \quad . \quad . \quad . \quad . \quad . \quad (13)$$

Hence, if U is the velocity of sound,

$$U^2 = \gamma p v. \quad . \quad . \quad . \quad . \quad . \quad . \quad . \quad . \quad . \quad . \quad . \quad (14)$$

We know that when the temperature is the same the product $p\,v$ remains constant. Hence, the velocity of sound is the same for the same temperature, whatever be the pressure of the air.

If $\mathfrak{H}$ is the height of the atmosphere supposed homogeneous—that is to say, the height of a column of the same density as the actual density, the weight of which would produce a pressure equal to the actual pressure—then, if the section of the column is unity, its volume is $\mathfrak{H}$, and if m is its mass, $\mathfrak{H} = m\,v$.

Also the weight of this column is $p = m\,g$, where g is the force of gravity.

Hence

$$p\,v = g\,\mathfrak{H}$$

and

$$U^2 = g\,\gamma\,\mathfrak{H}.$$

The velocity of sound may be compared with that of a body falling a certain distance under the action of gravity. For if V is the velocity of a body falling through a height s, $V^2 = 2\,g\,s$.

If we make $V = U$, then $s = \frac{1}{2}\,\gamma\,\mathfrak{H}$.

At the temperature of melting ice $\mathfrak{H} = 26{,}214$ feet if the force of gravity is 32·2.

At the same temperature the velocity of sound in air is 1,090 feet per second by experiment.

The square of this is 1,188,100, whereas the square of the velocity due to half the height of the homogeneous atmosphere is 843,821. Hence by means of the known velocity of sound we can determine γ, the ratio of 1,188,100 to 843,821, to be 1·408.

The height of the homogeneous atmosphere is proportional to the temperature reckoned from absolute zero. Hence the velocity of sound is proportional to the square root of the absolute temperature. In several of the more perfect gases the value of γ seems to be nearly the same as in air. Hence in those gases the velocity of sound is inversely as the square root of their specific gravity compared with air.

This investigation would be perfectly accurate, however great the changes of pressure and density due to the passage of the sound-wave, provided the substance is such that in the actual changes of pressure and volume the quantity

$$p + Q^2 v$$

remains constant, Q being the velocity of propagation. In all substances, as we have seen, we may, when the values of p and v are always very near their mean values, assume a value of Q which shall approximately satisfy this condition: but in the case of very violent sounds and other disturbances of the air the changes of p and v may be so great that this approximation ceases to be near the truth. To understand what takes place in these cases we must remember that the changes of p and v are not proportional to each other, for in almost all substances p increases faster for a given diminution of v as p increases and v diminishes.

Hence Q, which represents the mass of the substance traversed by the wave, will be greater in those parts of the wave where the pressure is great than in those parts where the pressure is small; that is, the condensed portions of the wave will travel faster than the rarefied portions. The result of this will be that if the wave originally consists of a gradual condensation followed by a gradual rarefaction, the condensation will become more sudden and the rarefaction more gradual as the wave advances through the air, in the same way and for nearly the same reason as the waves of the sea on coming into shallow water become steeper in front and more gently sloping behind, till at last they curl over on the shore.

FIG. 30.

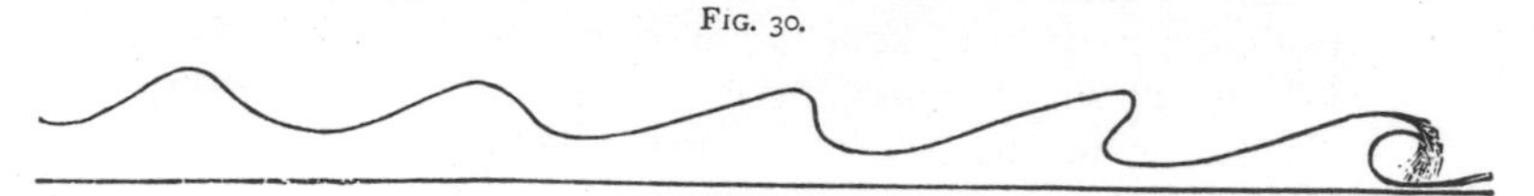

CHAPTER XVI.

ON RADIATION.

WE have already noticed some of the phenomena of radiation, and have shown that they do not properly belong to the science of Heat, and that they should rather be treated, along with sound and light, as a branch of the great science of Radiation.

The phenomenon of radiation consists in the transmission of energy from one body to another by propagation through the intervening medium, in such a way that the progress of the radiation may be traced, after it has left the first body and before it reaches the second, travelling through the medium with a certain velocity, and leaving the medium behind it in the condition in which it found it.

We have already considered one instance of radiation in the case of waves of sound. In this case the energy communicated to the air by a vibrating body is propagated through the air, and may finally set some other body, as the drum of the ear, in motion. During the propagation of the sound this energy exists in the portion of air through which it is travelling, partly in the form of motion of the air to and fro, and partly in the form of condensation and rarefaction. The energy due to sound in the air is distinct from heat, because it is propagated in a definite direction, so that in a certain time it will have entirely left the portion of air under consideration, and will be found in another portion of air to which it has travelled. Now heat never passes out of a hot body except to enter a colder body, so that the energy of sound-waves, or any other form of energy which is propagated so as to pass wholly out of one portion of the medium and into another, cannot be called heat.

There are, however, important thermal effects produced by radiation, so that we cannot understand the science of heat without studying some of the phenomena of radiation.

When a body is raised to a very high temperature it becomes visible in the dark, and is said to shine, or to emit light. The velocity of propagation of the light emitted by the sun and by very hot bodies has been approximately measured, and is estimated to be between 180,000 and 192,000 miles per second, or about 900,000 times faster than sound in air.

The time taken by the light in passing from one place to another within the limited range which we have at our command in a laboratory is exceedingly short, and it is only by means of the most refined experimental methods that it has been measured. It is certain, however, that there is an interval of time between the emission of light by one body and its reception by another, and that during this time the energy transmitted from the one body to the other has existed in some form in the intervening medium.

The opinions with regard to the relation between light and heat have suffered several alternations, according as these agents were regarded as substances or as accidents. At one time light was regarded as a substance projected from the luminous body, which, if the luminous body were hot, might itself become hot like any other substance. Heat was thus regarded as an accident of the substance light.

When the progress of science had rendered the measurement of quantities of heat as accurate as the measurement of quantities of gases, heat, under the name of caloric, was placed in the list of substances. Afterwards, the independent progress of optics led to the rejection of the corpuscular theory of light, and the establishment of the undulatory theory, according to which light is a wave-like motion of a medium already existing. The caloric theory of heat, however, still prevailed even after the corpuscular theory of light was rejected, so that heat and light seemed almost to have exchanged places.

When the caloric theory of heat was at length demonstrated to be false, the grounds of the argument were quite independent of those which had been used in the case of light.

We shall therefore consider the nature of radiation, whether of light or heat, in an independent manner, and show why we believe that what is called radiant heat is the same thing as what is called light, only perceived by us through a different channel. The same radiation which when we become aware of it by the eye we call light, when we detect it by a thermometer or by the sensation of heat we call radiant heat.

In the first place, radiant heat agrees with light in always moving in straight lines through any uniform medium. It is not, therefore, propagated by diffusion, as in the case of the conduction of heat, where the heat always travels from hotter to colder parts of the medium in whatever direction this condition may lead it.

The medium through which radiant heat passes is not heated if perfectly diathermanous, any more than a perfectly transparent medium through which light passes is rendered luminous. But if any impurity or defect of transparency causes the medium to become visible when light passes through it, it will also cause it to become hot and to stop part of the heat when traversed by radiant heat.

In the next place, radiant heat is reflected from the polished surfaces of bodies according to the same laws as light. A concave mirror collects the rays of the sun into a brilliantly luminous focus. If these collected rays fall on a piece of wood, they will set it on fire. If the luminous rays are collected by means of a convex lens, similar heating effects are produced, showing that radiant heat is refracted when it passes from one transparent medium to another.

When light is refracted through a prism, so as to change its direction through a considerable angle of deviation, it is separated into a series of kinds of light which are easily distinguished from each other by their various colours. The radiant heat which is refracted through the prism is also spread out through a considerable angular range, which shows that it also consists of radiations of various kinds. The luminosity of the different radiations is evidently not in the same proportion as their heating effects. For the blue and green rays have very little heating power compared with the extreme red, which are much less luminous, and the heating rays are found far beyond the end of the red, where no light at all is visible.

There are other methods of separating the different kinds of light, which are sometimes more convenient than the use of a prism. Many substances are more transparent to one kind of light than another, and are therefore called coloured media. Such media absorb certain rays and transmit others. If the light transmitted by a stratum of a coloured medium afterwards passes through another stratum of the same medium, it will be much less diminished in intensity than at first. For the kind of light which is most absorbed by the medium has been already removed, and what is transmitted by the first stratum is that which can pass most readily through the second. Thus a very thin stratum of a solution of bichromate of potash cuts off the whole of the spectrum from the middle of the green to the violet, but the remainder of the light, consisting of the red, orange, yellow, and part of the green, is very slightly diminished in intensity by passing through another stratum of the same medium.

If, however, the second stratum be of a different medium, which absorbs most of the rays which the first transmits, it will cut off nearly the whole light, though it may be itself very transparent for other rays absorbed by the first medium. Thus a stratum of sulphate of copper absorbs nearly all the rays transmitted by the bichromate of potash, except a few of the green rays.

Melloni found that different substances absorb different kinds of radiant heat, and that the heat sifted by a screen of any substance will pass in greater proportion through a screen of the same substance than unsifted heat, while it may be stopped in greater proportion than unsifted heat by a screen of a different substance.

These remarks may illustrate the general similarity between light and radiant heat. We must next consider the reasons which induce us to regard light as depending on a particular kind of motion in the medium through which it is propagated. These reasons are principally derived from the phenomena of the interference of light. They are explained more at large in treatises on light, because it is much easier to observe these phenomena by the eye than by any kind of thermometer. We shall therefore be as brief as possible.

There are various methods by which a beam of light from a small luminous object may be divided into two portions, which, after travelling by slightly different paths, finally fall on a white screen. Where the two portions of light overlap each other on the screen, a series of long narrow stripes may be seen, alternately lighter and darker than the average brightness of the screen near them, and when white light is used, these stripes are bordered with colours. By using light of one kind only, such as that obtained from the salted wick of a spirit-lamp, a greater number of bands or fringes may be seen, and a greater difference of brightness between the light and the dark bands. If we stop either of the portions of light into which the original beam was divided, the whole system of bands disappears, showing that they are due, not to either of the portions alone, but to both united.

If we now fix our attention on one of the dark bands, and then cut off one of the partial beams of light, we shall observe that instead of appearing darker it becomes actually brighter, and if we again allow the light to fall on the screen it becomes dark again. Hence it is possible to produce darkness by the addition of two portions of light. If light is a substance, there cannot be another substance which when added to it shall produce darkness. We are therefore compelled to admit that light is not a substance.

Now is there any other instance in which the addition of two apparently similar things diminishes the result? We know by experiments with musical instruments that a combination of two sounds may produce less audible effect than either separately, and it can be shown that this takes place when the one is half a wave-length in advance of the other. Here the mutual annihilation of the sounds arises from the fact that a motion of the air towards the ear is the exact opposite of a motion away from the ear, and if the two instruments are so arranged that the motions which they tend to produce in the air near the ear are in opposite directions and of equal magnitude, the result will be no motion at all. Now there is nothing absurd in one motion being the exact opposite of another, though the supposition that one substance is the exact opposite of another substance, as in some forms of the Two-Fluid theory of Electricity, is an absurdity.

We may show the interference of waves in a visible manner by dipping a two-pronged fork into water or mercury. The waves which diverge from the two centres where the prongs enter or leave the fluid are seen to produce a greater disturbance when they exactly coincide than when one gets ahead of the other.

Now it is found, by measuring the positions of the bright and dark bands on the screen, that the difference of the distances travelled by the two portions of light is for the bright bands always an exact multiple of a certain very small distance which we shall call a wave-length, whereas for the dark bands it is intermediate between two multiples of the wave-length, being $\frac{1}{2}$, $1\frac{1}{2}$, $2\frac{1}{2}$, &c., times that length.

We therefore conclude that whatever exists or takes place at a certain point in a ray of light, then, at the same instant, at a point at $\frac{1}{2}$ or $1\frac{1}{2}$ of the wave-length in advance, something exactly the opposite exists or takes place, so that in going along a ray we find an alternation of conditions which we may call positive and negative.

In the ordinary statement of the theory of undulations these conditions are described as motion of the medium in opposite directions. The essential character of the theory would remain the same if we were to substitute for ordinary motion to and fro any other succession of oppositely directed conditions. Professor Rankine has suggested opposite rotations of molecules about their axes, and I have suggested oppositely directed magnetizations and electromotive forces; but the adoption of either of these hypotheses would in no way alter the essential character of the undulatory theory.

Now it is found that if a very narrow thermo-electric pile be placed in the position of the screen, and moved so that sometimes a bright band and sometimes a dark one falls on the pile, the galvanometer indicates that the pile receives more heat when in the bright than when in the dark band, and that when one portion of the beam is cut off the heat in the dark band is increased. Hence in the interference of radiations the heating effect obeys the same laws as the luminous effect.

Indeed, it has been found that even when the source of radiation is a hot body which emits no luminous rays, the phenomena of interference can be traced, showing that two rays of dark heat can interfere no less than two rays of light. Hence all that we have said about the waves of light is applicable to the heat-radiation, which is therefore a series of waves.

It is also known in the case of light that after passing through a plate cut from a crystal of tourmaline parallel to its axis the transmitted beam cannot pass through a second similarly cut plate of tourmaline whose axis is perpendicular to that of the first, though it can pass through it when the axis is in any other position. Such a beam of light, which has different properties according as the second plate is turned into different positions round the beam as an axis, is called a polarized beam. There are many other ways of polarizing a beam of light, but the result is always of the same kind. Now this property of polarized light shows that the motion which constitutes light cannot be in the direction of the ray, for then there could be no difference between different sides of the ray. The motion must be transverse to the direction of the ray, so that we may now describe a ray of polarized light as a condition of disturbance in a direction at right angles to the ray propagated through a medium, so that the disturbance is in opposite directions at every half wave-length measured along the ray. Since Principal J. D. Forbes showed that a ray of dark heat can be polarized, we can make the same assertion about the heat radiation.

Let us now consider the consequences of admitting that what we call radiation, whether of heat, light, or invisible rays which act on chemical preparations, is of the nature of a transverse undulation in a medium.

A transverse undulation is completely defined when we know—

1. Its wave-length, or the distance between two places in which the disturbance is in the same phase.

2. Its amplitude, or the greatest extent of the disturbance.

3. The plane in which the direction of the disturbance lies.

4. The phase of the wave at a particular point.

5. The velocity of propagation through the medium.

When we know these particulars about an undulation, it is completely defined, and cannot be altered in any way without changing some of these specifications.

Now by passing a beam consisting of any assemblage of undulations through a prism, we can separate it into portions according to their wave-lengths, and we can select rays of a particular wave-length for examination. Of these we may, by means of a plate of tourmaline, select those whose plane of polarization is the principal plane of the tourmaline, but this is unnecessary for our purpose. We have now got rays of a definite wave-length. Their velocity of propagation depends only on the nature of the ray and of the medium, so that we cannot alter it at pleasure, and the phase changes so rapidly (billions of times in a second) that it cannot be directly observed. Hence the only variable quantity remaining is the amplitude of the disturbance, or, in other words, the intensity of the ray.

Now the ray may be observed in various ways. We may, if it excites the sensation of sight, receive it into our eye. If it affects chemical compounds, we may observe its effect on them, or we may receive the ray on a thermo-electric pile and determine its heating effect.

But all these effects, being effects of one and the same thing, must rise and fall together. A ray of specified wave-length and specified plane of polarization cannot be a combination of several different things, such as a light-ray, a heat-ray, and an actinic ray. It must be one and the same thing, which has luminous, thermal, and actinic effects, and everything which increases one of these effects must increase the others also.

The chief reason why so much that has been written on this subject is tainted with the notion that heat is one thing and light is another seems to be that the arrangements for operating on radiations of a selected wave-length are troublesome, and when mixed radiations are employed, in which the luminous and the thermal effects are in different proportions, anything which alters the proportion of the different radiations in the mixture alters also the proportion of the resulting thermal and luminous effect, as indeed it generally alters the colour of the mixed light.

We have seen that the existence of these radiations may be detected in various ways—by photographic preparations, by the eye, and by the thermometer. There can be no doubt, however, as to which of these methods gives the true measure of the energy transmitted by the radiation. This is exactly measured by the heating effect of the ray when completely absorbed by any substance.

When the wave-length is greater than 812 millionths of a millimetre no luminous effect is produced on the eye, though the effect on the thermometer may be very great. When the wave-length is 650 millionths of a millimetre the ray is visible as a red light, and a considerable heating effect is observed. But when the wave-length is 500 millionths of a millimetre, the ray, which is seen as a brilliant green, has much less heating effect than the dark or the red rays, and it is difficult to obtain strong thermal effects with rays of smaller wave-lengths, even when concentrated.

But, on the other hand, the photographic effect of the radiation on salts of silver, which is very feeble in the red rays, and even in the green rays, becomes more powerful the smaller the wave-length, till for rays whose wave-length is 400, which have a feeble violet luminosity and a still feebler thermal effect, the photographic effect is very powerful; and even far beyond the visible spectrum, for wave-lengths of less than 200 millionths of a millimetre, which are quite invisible to our eyes and quite undiscoverable by our thermometers, the photographic effect is still observed. This shows that neither the luminous nor the photographic effect is in any way proportional to the energy of the radiation when different kinds of radiation are concerned. It is probable that when the radiation produces the photographic effect it is not by its energy doing work on the chemical compound, but rather by a well-timed vibration of the molecules dislodging them from the position of almost indifferent equilibrium into which they had been thrown by previous chemical manipulations, and enabling them to rush together according to their more permanent affinities, so as to form stabler compounds. In cases of this kind the effect is no more a dynamical measure of the cause than the effect of the fall of a tree is a measure of the energy of the wind which uprooted it.

It is true that in many cases the amount of the radiation may be very accurately estimated by means of its chemical effects, even when these chemical effects tend to diminish the intrinsic energy of the system. But by estimating the heating effect of a radiation which is entirely absorbed by the heated body we obtain a true measure of the energy of the radiation. It is found that a surface thickly coated with lampblack absorbs nearly the whole of every kind of radiation which falls on it. Hence surfaces of this kind are of great value in the thermal study of radiation.

We have now to consider the conditions which determine the amount and quality of the radiation from a heated body. We must bear in mind that temperature is a property of hot bodies and not of radiations, and that qualities such as wave-lengths, &c., belong to radiations, but not to the heat which produces them or is produced by them.

ON PREVOST'S THEORY OF EXCHANGES.

When a system of bodies at different temperatures is left to itself, the transfer of heat which takes place always has the effect of rendering the temperatures of the different bodies more nearly equal, and this character of the transfer of heat, that it passes from hotter to colder bodies, is the same whether it is by radiation or by conduction that the transfer takes place.

Let us consider a number of bodies, all at the same temperature, placed in a chamber the walls of which are maintained at that temperature, and through which no heat can pass by radiation (suppose the walls of metal, for instance). No change of temperature will occur in any of these bodies. They will be in thermal equilibrium with each other and with the walls of the chamber. This is a consequence of the definition of equal temperature at p. 32.

Now if any one of these bodies had been taken out of the chamber and placed among colder bodies there would be a transfer of heat by radiation from the hot body to the colder ones; or if a colder body had been introduced into the chamber it would immediately begin to receive heat by radiation from the hotter bodies round it. But the cold body has no power of acting directly on the hot bodies at a distance, so as to cause them to begin to emit radiations, nor has the hot chamber any power to stop the radiation of any one of the hot bodies placed within it. We therefore conclude with Prevost that a hot body is always emitting radiations, even when no colder body is there to receive them, and that the reason why there is no change of temperature when a body is placed in a chamber of the same temperature is that it receives from the radiation of the walls of the chamber exactly as much heat as it loses by radiation towards these walls.

If this is the true explanation of the thermal equilibrium of radiation, it follows that if two bodies have the same temperature the radiation emitted by the first and absorbed by the second is equal in amount to the radiation emitted by the second and absorbed by the first during the same time.

The higher the temperature of a body, the greater its radiation is found to be, so that when the temperatures of the bodies are unequal the hotter bodies will emit more radiation than they receive from the colder bodies, and therefore, on the whole, heat will be lost by the hotter and gained by the colder bodies till thermal equilibrium is attained. We shall return to the comparison of the radiation at different temperatures after we have examined the relations between the radiation of different bodies at the same temperature.

The application of the theory of exchanges has at various times been extended to the phenomena of heat as they were successively investigated Fourier has considered the law of radiation as depending on the angle which the ray makes with the surface, and Leslie has investigated its relation to the state of polish of the surface; but it is in recent times, and chiefly by the researches of B. Stewart, Kirchhoff, and De la Provostaye, that the theory of exchanges has been shown to be applicable, not only to the total amount of the radiation, but to every distinction in quality of which the radiation is capable.

For, by placing between two bodies of the same temperature a contrivance such as that already noticed at p. 218, so that only radiations of a determinate wave-length and in a determinate plane can pass from the one body to the other, we reduce the general proposition about thermal equilibrium to a proposition about this particular kind of radiation. We may therefore transform it into the following more definite proposition.

If two bodies are at the same temperature, the radiation emitted by the first and absorbed by the second agrees with the radiation emitted by the second and absorbed by the first, not only in its total heating effect, but in the intensity, wave-length, and plane of polarization of every component part of either radiation. And the law that the amount of radiation increases with the temperature must be true, not only for the whole radiation, but for all the component parts of it when analysed according to their wave-lengths and planes of polarization.

The consequences of these two propositions, applying as they do to every kind of radiation, whether detected by its thermal or by its luminous effects, are so numerous and varied that we cannot attempt any full enumeration of them in this treatise. We must confine ourselves to a few examples.

When a radiation falls on a body, part of it is reflected, and part enters the body. The latter part again may either be wholly absorbed by the body or partly absorbed and partly transmitted.

Now lampblack reflects hardly any of the radiation which falls on it, and it transmits none. Nearly the whole is absorbed.

Polished silver reflects nearly the whole of the radiation which falls upon it, absorbing only about a fortieth part, and transmitting none.

Rock salt reflects less than a twelfth part of the radiation which falls on it; it absorbs hardly any, and transmits ninety-two per cent.

These three substances, therefore, may be taken as types of absorption, reflexion, and transmission respectively.

Let us suppose that these properties have been observed in these substances at the temperature, say, of 212° F., and let them be placed at this temperature within a chamber whose walls are at the same temperature. Then the amount of the radiation from the lampblack which is absorbed by the other two substances is, as we have seen, very small. Now the lampblack absorbs the whole of the radiation from the silver or the salt. Hence the radiation from these substances must also be small, or, more precisely—

The radiation of a substance at a given temperature is to the radiation of lampblack at that temperature as the amount of radiation absorbed by the substance at that temperature is to the whole radiation which falls upon it.

Hence a body whose surface is made of polished silver will emit a much smaller amount of radiation than one whose surface is of lampblack. The brighter the surface of a silver teapot, the longer will it retain the heat of the tea; and if on the surface of a metal plate some parts are polished, others rough, and others blackened, when the plate is made red hot the blackened parts will appear brightest, the rough parts not so bright, and the polished parts darkest. This is well seen when melted lead is made red hot. When part of the dross is removed, the polished surface of the melted metal, though really hotter than the dross, appears of a less brilliant red.

A piece of glass when taken red hot out of the fire appears of a very faint red compared with a piece of iron taken from the same part of the fire, though the glass is really hotter than the iron, because it does not throw off its heat so fast.

Air or any other transparent gas, even when raised to a heat at which opaque bodies appear white hot, emits so little light that its luminosity can hardly be observed in the dark, at least when the thickness of the heated air is not very great.

Again, when a substance at a given temperature absorbs certain kinds of radiation and transmits others, it emits at that temperature only those kinds of radiation which it absorbs. A very remarkable instance of this is observed in the vapour of sodium. This substance when heated emits rays of two definite kinds, whose wave-lengths are 0·00059053 and 0·00058989 millimetre respectively. These rays are visible, and may be seen in the form of two bright lines by directing a spectroscope upon a flame in which any compound of sodium is present.

Now if the light emitted from an intensely heated solid body, such as a piece of lime in the oxyhydrogen light, be transmitted through sodium-vapour at a temperature lower than that of the lime, and then analysed by the spectroscope, two dark lines are seen, corresponding to the two bright ones formerly observed, showing that sodium-vapour absorbs the same definite kinds of light which it radiates.

If the temperature of the sodium-vapour is raised, say by using a Bunsen's burner instead of a spirit-lamp to produce it, or if the temperature of the lime is lowered till it is the same as that of the vapour, the dark lines disappear, because the sodium-vapour now radiates exactly as much light as it absorbs from the light of the lime-ball at the same temperature. If the sodium-flame is hotter than the lime-ball the lines appear bright.

This is an illustration of Kirchhoff's principle, that the radiation of every kind increases as the temperature rises.

In performing this experiment we suppose the light from the lime-ball to pass through the sodium-flame before it reaches the slit of the spectroscope. If, however, the flame is interposed between the slit and the eye, or the screen on which the spectrum is projected, the dark lines may be seen distinctly, even when the temperature of the sodium-flame is higher than that of the lime-ball. For in the parts of the spectrum near the lines the light is now compounded of the analysed light of the lime-ball and the direct light of the sodium-flame, while at the lines themselves the light of the spectrum of the lime-ball is cut off, and only the direct light of the sodium-flame remains, so that the lines appear darker than the rest of the field.

It does not belong to the scope of this treatise to attempt to go over the immense field of research which has been opened up by the application of the spectroscope to distinguish different incandescent vapours, and which has led to a great increase of our knowledge of the heavenly bodies.

If the thickness of a medium, such as sodium-vapour, which radiates and absorbs definite kinds of light, be very great, the whole being at a high temperature, the light emitted will be of exactly the same composition as that emitted from lampblack at the same temperature. For, though some kinds of radiation are much more feebly emitted by the substance than others, these are also so feebly absorbed that they can reach the surface from immense depths, whereas the rays which are so copiously radiated are also so rapidly absorbed that it is only from places very near the surface that they can escape out of the medium. Hence both the depth and the density of an incandescent gas cause its radiation to assume more and more of the character of a continuous spectrum.

When the temperature of a substance is gradually raised, not only does the intensity of every particular kind of radiation increase, but new kinds of radiation are produced. Bodies of low temperature emit only rays of great wave-length. As the temperature rises these rays are more copiously emitted, but at the same time other rays of smaller wave-length make their appearance. When the temperature has risen to a certain point, part of the radiation is luminous and of a red colour, the luminous rays of greatest wave-length being red. As the temperature rises, the other luminous rays appear in the order of the spectrum, but every rise of temperature increases the intensity of all the rays which have already made their appearance. A white-hot body emits more red rays than a red-hot body, and more non-luminous rays than any non-luminous body.

The total thermal value of the radiation at any temperature, depending as it does upon the amount of all the different kinds of rays of which it is composed, is not likely to be a simple function of the temperature. Nevertheless, Dulong and Petit succeeded in obtaining a formula which expresses the facts observed by them with tolerable exactness. It is of the form

$$R = m\,a^{\theta},$$

where R is the total loss of heat in unit of time by radiation from unit of area of the surface of the substance at the temperature θ, m is a constant quantity depending only on the substance and the nature of its surface, and a is a numerical quantity which, when θ expresses the temperature on the Centigrade scale, is 1·0077.

If the body is placed in a chamber devoid of air, whose walls are at the temperature t, then the heat radiated from the walls to the body and absorbed by it will be

$$r = m\,a^{t},$$

so that the actual loss of heat will be

$$\mathrm{R} - r = m\,a^{\theta} - m\,a^{t}.$$

The constancy of the amount of radiation between the same surfaces at the same temperatures affords a very convenient method of comparing quantities of heat. This method was referred to in our chapter on Calorimetry (p. 74), under the name of the Method of Cooling.

The substance to be examined is heated and put into a thin copper vessel, the outer surface of which is blackened, or at least is preserved in the same state of roughness or of polish throughout the experiments. This vessel is placed in a larger copper vessel so as not to touch it, and the outer vessel is placed in a bath of water kept at a constant temperature. The temperature of the substance in the smaller vessel is observed from time to time, or, still better, the times are observed at which the reading of a thermometer immersed in the substance is an exact number of degrees. In this way the time of cooling, say from 100° to 90°, from 90° to 80°, is registered, the temperature of the outer vessel being kept always the same.

Suppose that this observation of the time of cooling is made first when the vessel is filled with water, and then when some other substance is put into it. The rate at which heat escapes by radiation is the same for the same temperature in both experiments. The quantity of heat which escapes during the cooling, say from 100° to 90°, in the two experiments, is proportional to the time of cooling. Hence the capacity of the vessel and its contents in the first experiment is to its capacity in the second experiment as the time fcooling from 100° to 90° in the first experiment is to the time of cooling from 100° to 90° in the second experiment.

The method of cooling is very convenient in certain cases, but it is necessary to keep the temperature of the whole of the substance in the inner vessel as nearly uniform as possible, so that the method must be restricted to liquids which we can stir, and to solids whose conductivity is great, and which may be cut in pieces and immersed in a liquid.

The method of cooling has been found very applicable to the measurement of the quantity of heat conducted through a substance. (See the chapter on Conduction.)

EFFECT OF RADIATION ON THERMOMETERS.

On account of the radiation passing in all directions through the atmosphere, it is a very difficult thing to determine the true temperature of the air in any place out of doors by means of a thermometer.

If the sun shines on the thermometer, the reading is of course too high; but if we put it in the shade, it may be too low, because the thermometer may be emitting more radiation than it receives from the clear sky. The ground, walls of houses, clouds, and the various devices for shielding the thermometer from radiation, may all become sources of error, by causing an unknown amount of radiation on the bulb. For rough purposes the effects of radiation may be greatly removed by giving the bulb a surface of polished silver, of which, as we have seen, the absorption is only a fortieth of that of lampblack.

A method described by Dr. Joule in a communication to the Philosophical Society of Manchester, November 26, 1867, seems the only one free from all objections. The thermometer is placed in a long vertical copper tube open at both ends, but with a cap to close the lower end, which may be removed or put on without warming it by the hand. Whatever radiation affects the thermometer must be between it and the inside of the tube, and if these are of the same temperature, the radiation will have no effect on the observed reading of the thermometer. Hence, if we can be sure that the copper tube and the air within it are at the temperature of the atmosphere, and that the thermometer is in thermal equilibrium, the thermometer reading will be the true temperature.

Now, if the air within the tube is of the same temperature as the air outside, it will be of the same density, and it will therefore be in statical equilibrium with it. If it is warmer it will be lighter, and an upward current will be formed in the tube when the cap is removed. If it is colder, a downward current will be formed.

To detect these currents a spiral wire is suspended in the tube by a fine fibre, so that an upward or downward current causes the spiral to twist the fibre, and any motion of the spiral is made apparent by means of a small mirror attached to it.

To vary the temperature of the copper tube, it is enclosed in a wider tube, so that water may be placed in the space between the tubes, and by pouring in warmer or cooler water the temperature may be adjusted till there is no current.

We then know that the air is of the same temperature within the tube as it is without. But we know that the tube is also of the same temperature as the air, for if it were not it would heat or cool the air and produce a current. Finally, we know that the thermometer, if stationary, is at the temperature of the atmosphere; for the air in contact with it, and the sides of the tube, which alone can exchange radiations with it, have the same temperature as the atmosphere.

CHAPTER XVII.

ON CONVECTION CURRENTS.

WHEN the application of heat to a fluid causes it to expand or to contract, it is thereby rendered rarer or denser than the neighbouring parts of the fluid ; and if the fluid is at the same time acted on by gravity, it tends to form an upward or downward current of the heated fluid, which is of course accompanied with a current of the more remote parts of the fluid in the opposite direction. The fluid is thus made to circulate, fresh portions of fluid are brought into the neighbourhood of the source of heat, and these when heated travel, carrying their heat with them into other regions. Such currents, caused by the application of heat, and carrying this heat with them, are called convection currents. They play a most important part in natural phenomena, by causing a much more rapid diffusion of heat than would take place by conduction alone in the same medium if restrained from moving. The actual diffusion of heat from one part of the fluid to another takes place, of course, by conduction ; but, on account of the motion of the fluid, the isothermal surfaces are so extended, and in some cases contorted, that their areas are greatly increased while the distances between them are diminished, so that true conduction goes on much more rapidly than if the medium were at rest.

Convection currents depend on changes of density in a fluid acted on by gravity. If the action of heat does not produce a change of density, as in the case of water at a temperature of about 39° F., no convection current will be produced. If the fluid is not acted on by gravity, as would be the case if the fluid were removed to a sufficient distance from the earth and other great bodies, no convection currents would be formed. As this condition is not easily realised, we may take the case of a vessel containing fluid, and descending according to the law of motion of a body falling freely. The pressure in this fluid will be the same in every part, and a change of density in any part of the fluid will not occasion convection currents.

When we wish to avoid the formation of convection currents we must arrange matters so that during the whole course of the experiment the density of each horizontal stratum is the same throughout, and that the density increases with the depth. If, for instance, we are studying the conduction of heat in a fluid which expands when heated, we must make the heat flow downwards through the fluid. If we wish to determine the law of diffusion of fluids we must place the denser fluid underneath the rarer one.

Convection currents are produced by changes of density arising from other causes. Thus if a crystal of a soluble salt be suspended in a vessel of water, the water in contact with the crystal will dissolve a portion of it, and, becoming denser, will begin to sink, and its place will be supplied by fresh water. Thus a convection current will be formed, a solution of the salt will descend from the crystal, and this will cause an upward current of purer water, and a circulation will be kept up till either the crystal is entirely dissolved, or the liquid has become saturated with the salt up to the level of the top of the crystal. In this case it is the salt which is carried through the liquid by convection.

A convection current may be produced in which electricity is the thing carried. If a conductor terminating in a fine point is strongly electrified, the particles of air near the point will be charged with electricity, and then urged from the point towards any surface oppositely electrified. A current of electrified air is thus formed, which diffuses itself about the room, and generally reaches the walls, where the electrified air clings to the oppositely electrified wall, and is sometimes not discharged for a long time.

The method of determining by convection currents the temperature at which water has its maximum density seems to have been first employed by Hope. He cooled the middle part of a tall vessel of water by surrounding this part of the vessel with a freezing mixture. As long as the temperature is above 40° F. the cooled water descends, and causes a fall of temperature in a thermometer placed in the lower part of the vessel. Another thermometer, placed in the upper part of the vessel, remains stationary. But when the temperature is below 39° F. the water cooled by the freezing mixture becomes lighter and ascends, causing the upper thermometer to fall, while the lower one remains stationary.

The investigation of the maximum density of water has been greatly improved by Joule, who also made use of convection currents. He employed a vessel consisting of two vertical cylinders, each $4\frac{1}{2}$ feet high and 6 inches diameter, connected below by a wide tube with a cock, and above by an open trough or channel. The whole was filled with water up to such a level that the water could flow freely through the channel. A glass specific gravity bead which would just float in water was placed in the channel, and served to indicate any motion of the water in the channel. The very smallest difference of density between the portions of water in the two columns was sufficient to produce a current, and to move the bead in the channel.

FIG. 31.

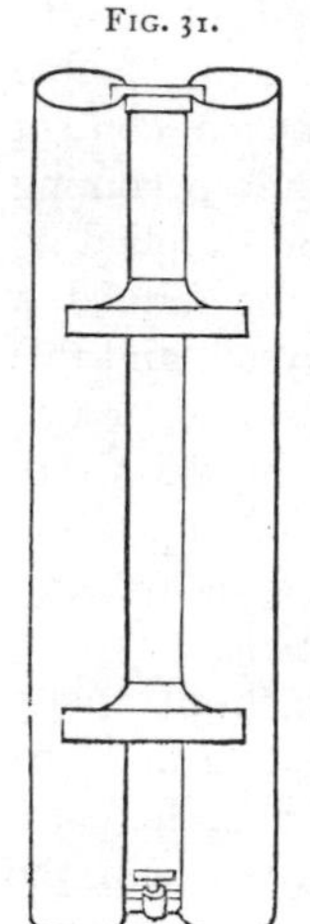

The cock in the connecting tube being closed, the temperature of the water in the two tubes was adjusted, the water well mixed in each tube by stirring, and when it had come to rest the temperature of each column was observed, and the cock was opened. If a current was then observed in the channel, it indicated that the water in the tube towards which the current flowed was the denser. By finding a pair of different temperatures at which the density is exactly the same, we may be sure that one of them is below and the other above the temperature of maximum density; and by obtaining a series of such pairs of temperatures of which the difference is smaller and smaller, Dr. Joule determined the temperature of maximum density to be 39°·1 F. within a very small fraction of a degree.

CHAPTER XVIII.

ON THE DIFFUSION OF HEAT BY CONDUCTION.

WHENEVER different parts of a body are at different temperatures, heat flows from the hotter parts to the neighbouring colder parts. To obtain an exact notion of conduction, let us consider a large boiler with a flat bottom, whose thickness is c. The fire maintains the lower surface at the temperature T, and heat flows upwards through the boiler plate to the upper surface, which is in contact with the water at the lower temperature, S.

FIG. 32.

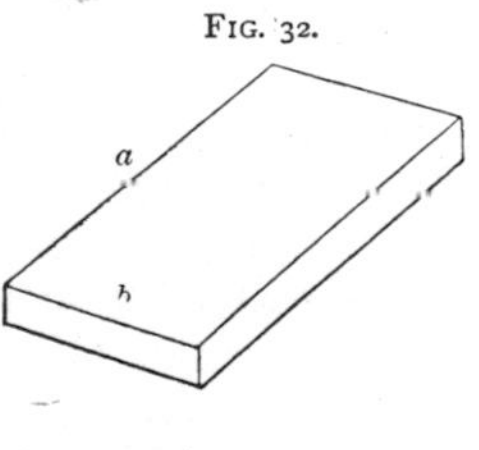

Let us now restrict ourselves to the consideration of a rectangular portion of the boiler plate, whose length is a, its breadth b, and its thickness c.

The things to be considered are the dimensions of this portion of the body, and the nature of the material of which it is made, the temperatures of its upper and lower surfaces, and the flow of heat through it as determined by these conditions. In the first place it is found that when the difference of the temperatures S and T is not so great as to make a sensible difference between the properties of the substance at these two temperatures, the flow of heat is exactly proportional to the difference of temperatures, other things being the same.

Let us suppose that when a, b, and c are each equal to the unit of length, and when T is one degree above S, the steady flow of heat is such that the quantity which enters the lower surface or leaves the upper surface in the unit of time is k, then k is defined as the specific thermal conductivity of the substance. To find H, the quantity of heat which flows in a time t through the portion of boiler plate whose area is $a\,b$, and whose thickness is c, when the lower surface is kept at a temperature T, and the upper at a temperature S, till the flow has become steady, divide the plate into c horizontal layers, the thickness of each layer being unity, and divide each layer into $a\,b$ cubes, the sides of each cube being unity.

Since the flow of heat is steady, the difference of temperature of the upper and lower faces of each cube will be $\frac{1}{c}$ (T — S). The flow of heat through each cube will be $\frac{k}{c}$ (T—S) in unit of time. Now, in each layer there are $a\,b$ such cubes, and the flow goes on for t units of time, so that we obtain for the whole heat conducted in time t

$$\text{H} = \frac{a\,b\,t\,k}{c}\ (\text{T} - \text{S}),$$

where $a\,b$ is the area and c the thickness of the plate, t the time, T — S the difference of temperature which causes the flow, and k the specific thermal conductivity of the substance of the plate.

It appears, therefore, that the heat conducted is directly proportional to the area of the plate, to the time, to the difference of temperature, and to the conductivity, and inversely proportional to the thickness of the plate.

ON THE DIMENSIONS OF k, THE SPECIFIC THERMAL CONDUCTIVITY.

From the equation we find

$$k = \frac{c\,\mathrm{H}}{a\,b\,t\,(\mathrm{T}-\mathrm{S})}.$$

Hence if $[\mathrm{L}]$ be the unit of length, $[\mathrm{T}]$ the unit of time, $[\mathrm{H}]$ the unit of heat, and $[\Theta]$ the unit of temperature, the dimensions of k will be $\frac{[\mathrm{H}]}{[\mathrm{L\,T}\Theta]}$.

The further discussion of the dimensions of k will depend on the mode of measuring heat and temperature.

(1) If heat is measured as energy, its dimensions are $\left[\frac{\mathrm{L}^2\,\mathrm{M}}{\mathrm{T}^2}\right]$, and those of k become $\left[\frac{\mathrm{L\,M}}{\mathrm{T}^3\,\Theta}\right]$. This may be called the *dynamical* measure of the conductivity.

(2) If heat is measured in thermal units, such that each thermal unit is capable of raising unit of mass of a standard substance through one degree of temperature, the dimensions of H are $[\mathrm{M}\,\Theta]$, and those of k will be $\left[\frac{\mathrm{M}}{\mathrm{L\,T}}\right]$. This may be called the *calorimetric* measure of the conductivity.

(3) If we take as the unit of heat that which will raise unit of *volume* of the substance itself one degree, the dimensions of H are $[\mathrm{L}^3\Theta]$, and those of k are $\left[\frac{\mathrm{L}^2}{\mathrm{T}}\right]$. This may be called the *thermometric* measure of the conductivity.

In order to obtain a distinct conception of the flow of heat through a solid body, let us suppose that at a given instant we know the temperature of every point of the body. If we now suppose a surface to be described within the body such that at every point of this surface the temperature has a given value T°, we may call this surface the isothermal surface of T°. (Of course, when we suppose this surface to exist in the body, we do not conceive the body to be altered in any way by this supposition, as if the body were really cut in two by it.) This isothermal surface separates those parts of the body which are hotter than the temperature T° from those which are colder than this temperature.

Let us now suppose the isothermal surfaces drawn for every exact degree of temperature, from that of the hottest part of the body to that of the coldest part. These surfaces may be curved in any way, but no two different surfaces can meet each other, because no part of the body can at the same time have two different temperatures. The body will therefore be divided into layers or coats by these surfaces, and the space between two isothermal surfaces differing by one degree of temperature will be in the form of a thin shell, whose thickness may vary from one part to another.

At every point of this shell there is a flow of heat from the hotter surface to the colder surface through the substance of the shell.

The direction of this flow is perpendicular to the surface of the shell, and the rate of flow is greater the thinner the shell is at the place, and the greater its conductivity.

If we draw a line perpendicular to the surface of the shell, and of length unity, then if c is the thickness of the shell, and if the neighbouring shells are of nearly the same thickness, this line will cut a number of shells equal to $\frac{1}{c}$. This, then, is the difference of temperature between two points in the body at unit of distance, measured in the direction of the flow of heat, and therefore the flow of heat along this line is measured by $\frac{k}{c}$, where k is the conductivity.

We can now imagine, with the help of the isothermal surfaces, the state of the body at a given instant. Wherever there is inequality of temperature between neighbouring parts of the body a flow of heat is going on. This flow is everywhere perpendicular to the isothermal surfaces, and the flow through unit of area of one of these surfaces in unit of time is equal to the conductivity divided by the distance between two consecutive isothermal surfaces.

The knowledge of the actual thermal state of the body, and of the law of conduction of heat, thus enables us to determine the flow of heat at every part of the body. If the flow of heat is such that the amount of heat which flows into any portion of the body is exactly equal to that which flows out of it, then the thermal state of this portion of the body will remain the same as long as the flow of heat fulfils this condition.

If this condition is fulfilled for every part of the body, the temperature at any point will not alter with the time, the system of isothermal surfaces will continue the same, and the flow of heat will go on without alteration, being always the same at the same part of the body.

This state of things is referred to as the *state of steady flow of heat*. It cannot exist unless heat is steadily supplied to the hotter parts of the surface of the body, from some source external to the body, and an equal quantity removed from the colder parts of the surface by some cooling medium, or by radiation.

The state of steady flow of heat requires the fulfilment at every part of the body of a certain condition, similar to that which is fulfilled in the flow of an incompressible fluid.

When this condition is not fulfilled, the quantity of heat which enters any portion of the body may be greater or less than that which escapes from it. In the one case heat will accumulate, and the portion of the body will rise in temperature. In the other case the heat of the portion will diminish, and it will fall in temperature. The amount of this rise or fall of temperature will be measured numerically by the gain or loss of heat, divided by the capacity for heat of the portion considered.

If the portion considered is unit of volume, and if we measure heat as in the third method given at p. 235 by the quantity required to raise unit of volume of the substance, in its actual state, one degree, then the rise of temperature of this portion will be numerically equal to the total flow of heat into it.

We are now able, by means of a thorough knowledge of the thermal state of the body at a given instant, to determine the rate at which the temperature of every part must be changing, and therefore we are able to predict its state in the succeeding instant. Knowing this, we can predict its state in the next instant following, and so on.

The only parts of the body to which this method does not apply are those parts of its surface to which heat is supplied, or from which heat is abstracted, by agencies external to the body. If we know either the rate at which heat is supplied or abstracted at every part of the surface, or the actual temperature of every part of the surface during the whole time, either of these conditions, together with the original thermal state of the body, will afford sufficient data for calculating the temperature of every point during all time to come.

The discussion of this problem is the subject of the great work of Joseph Fourier, *Théorie de la Chaleur*. It is not possible in a treatise of the size and scope of this book to reproduce, or even to explain, the powerful analytical methods employed by Fourier to express the varied conditions, as to the form of its surface and its original thermal state, to which the body may be subjected. These methods belong, rather, to the general theory of the application of mathematics to physics; for in every branch of physics, when the investigation turns upon the expression of arbitrary conditions, we have to follow the method which Fourier first pointed out in his 'Theory of Heat.'

I shall only mention one or two of the results given by Fourier, in which the intricacies arising from the arbitrary conditions of the problem are avoided.

The first of these is the case in which the solid is supposed of infinite extent, and of the same conductivity in every part.

The temperature of every point of this body at a given time is supposed to be known, and it is required to determine the temperature of any given point P after a time t has elapsed.

Fourier has given a complete solution of this problem, of which we may obtain some idea by means of the following considerations. Let k be the conductivity, measured by the third method, in which the unit of heat adopted is that which will raise unit of volume of the substance one degree; then if we make

$$k\,t = a^2,$$

a will be a line the length of which will be proportional to the square root of the time.

Let Q be any point in the body, and let its distance from P be r. Let the original temperature of Q be θ. Now take a quantity of matter proportional to $e^{-\frac{r^2}{4kt}}$ and of the temperature θ, and mix it with portions of matter taken from every other part of the body, the temperature of each portion being the original temperature of that point, and the quantity of each portion being proportional to $e^{-\frac{r^2}{4kt}}$. The mean temperature of all such portions will be the temperature of the point P after a time t.

In other words, the temperature of P after a time t may be regarded as in some sense the mean of the original temperatures of all parts of the body. In taking this mean, however, different parts are allowed different weights, depending on their distance from P, the parts near P having more influence on the result than those at a greater distance.

The mathematical formula which indicates the weight to be given to the temperature of each part in taking the mean is a very important one. It occurs in several branches of physics, particularly in the theory of errors and in that of the motions of systems of molecules.

It follows from this result that, in calculating the temperature of the point P, we must take into account the temperature of every other point Q, however distant, and however short the time may be during which the propagation of heat has been going on. Hence, in a strict sense, the influence of a heated part of the body extends to the most distant parts of the body in an incalculably short time, so that it is impossible to assign to the propagation of heat a definite velocity. The velocity of propagation of thermal effects depends entirely on the magnitude of the effect which we are able to recognise; and if there were no limit to the sensibility of our instruments, there would be no limit to the rapidity with which we could detect the influence of heat applied to distant parts of the body. But while this influence on distant points can be expressed mathematically from the first instant, its numerical value is excessively small until, by the lapse of time, the line a has grown so as to be comparable with r, the distance of P from Q. If we take this into consideration, and remember that it is only when the changes of temperature are comparable with the original differences of temperature that we can detect them with our instruments, we shall see that the sensible propagation of heat, so far from being instantaneous, is an excessively slow process, and that the time required to produce a similar change of temperature in two similar systems of different dimensions is proportional to the *square* of the linear dimensions. For instance, if a red-hot ball of four inches' diameter fired into a sandbank has in an hour raised the temperature of the sand six inches from its centre 10° F., then a red-hot ball of eight inches' diameter would take four hours to raise the temperature of the sand twelve inches from its centre by the same number of degrees.

This result, which is very important in practical questions about the time of cooling or heating of bodies of any form, may be deduced directly from the consideration of the dimensions of the quantity k—namely, the square of a length divided by a time. It follows from this that if in two unequally heated systems of similar form but different dimensions the conductivity and the temperature are the same at corresponding points at first, then the process of diffusion of heat will go on at different rates in the two systems, so that if for each system the time be taken proportional to the square of the linear dimensions, the temperatures of corresponding points will still be the same in both systems.

The method just described affords a complete determination of the temperature of any point of a homogeneous infinite solid at any future time, the temperature of every point of the solid being given at the instant from which we begin to count the time. But when we attempt to deduce from a knowledge of the present thermal state of the body what must have been its state at some past time, we find that the method ceases to be applicable.

To make this attempt, we have only to make t, the symbol of the time, a negative quantity in the expressions given by Fourier. If we adopt the method of taking the mean of the temperatures of all the particles of the solid, each particle having a certain weight assigned to it in taking the mean, we find that this weight, according to the formula, is greater for the distant particles than for the neighbouring ones, a result sufficiently startling in itself. But when we find that, in order to obtain the mean, after taking the sum of the temperatures multiplied by their proper factors, we have to divide by a quantity involving the square root of t, the time, we are assured that when t is taken negative the operation is simply impossible, and devoid of any physical meaning, for the square root of a negative quantity, though it may be interpreted with reference to some geometrical operations, is absolutely without meaning with reference to time.

It appears, therefore, that Fourier's solution of this

problem, though complete considered with reference to future time, fails when we attempt to discover the state of the body in past time.

In the diagram fig. 33 the curves show the distribution of

FIG. 33.

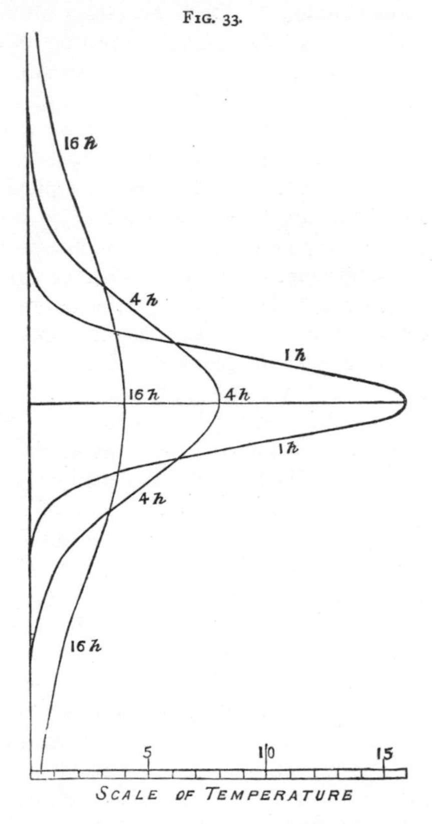

temperature in an infinite mass at different times, after the sudden introduction of a hot horizontal stratum in the midst of the infinite solid. The temperature is indicated by the horizontal distance to the right of the vertical line, and the hot stratum is supposed to have been introduced at the middle of the figure.

The curves indicate the temperatures of the various strata one hour, four hours, and sixteen hours after the introduction of the hot stratum. The gradual diffusion of the heat is evident, and also the diminishing rate of diffusion as its extent increases.

The problem of the diffusion of heat in an infinite solid does not present those difficulties which occur in problems relating to a solid of definite shape. These difficulties arise from the conditions to which the surface of the solid may be subjected, as, for instance, the temperature may be given over part of the surface, the quantity of heat supplied to another part may be given, or we may only know that the surface is exposed to air of a certain temperature.

The method by which Fourier was enabled to solve many questions of this kind depends on the discovery of harmonic distributions of heat.

Suppose the temperatures of the different parts of the body to be so adjusted that when the body is left to itself under the given conditions relating to the surface, the temperatures of all the parts converge to the final temperature, their differences from the final temperature always preserving the same proportion during the process; then this distribution of temperature is called an harmonic distribution. If we suppose the final temperature to be taken as zero, then the temperatures in the harmonic distribution diminish in a geometrical progression as the times increase in arithmetical progression, the ratio of cooling being the same for all parts of the body.

In each of the cases investigated by Fourier there may be an infinite series of harmonic distributions. One of these, which has the slowest rate of diminution, may be called the fundamental harmonic; the rates of diminution of the others are proportional to the squares of the natural numbers.

If the body is originally heated in any arbitrary manner, Fourier shows how to express the original temperature as the sum of a series of harmonic distributions. When the body is left to itself the part depending on the higher harmonics rapidly dies away, so that after a certain time the distribution of heat continually approximates to that due to the fundamental harmonic, which therefore represents the law of cooling of a body after the process of diffusion of heat has gone on for a long time.

Sir William Thomson has shown, in a paper published in the 'Cambridge and Dublin Mathematical Journal' in 1844, how to deduce, in certain cases, the thermal state of a body in past time from its observed condition at present.

For this purpose, the present distribution of temperature must be expressed (as it always may be) as the sum of a series of harmonic distributions. Each of these harmonic distributions is such that the difference of the temperature of any point from the final temperature diminishes in a geometrical progression as the time increases in arithmetical progression, the ratio of the geometrical progression being the greater the higher the degree of the harmonic.

If we now make t negative, and trace the history of the distribution of temperature up the stream of time, we shall find each harmonic increasing as we go backwards, and the higher harmonics increasing faster than the lower ones.

If the present distribution of temperature is such that it may be expressed in a finite series of harmonics, the distribution of temperature at any previous time may be calculated; but if (as is generally the case) the series of harmonics is infinite, then the temperature can be calculated only when this series is convergent. For present and future time it is always convergent, but for past time it becomes ultimately divergent when the time is taken at a sufficiently remote epoch. The negative value of t, for which the series becomes ultimately divergent, indicates a certain date in past time such that the present state of things cannot be deduced from any distribution of temperature occurring previously to that date, and becoming diffused by ordinary conduction. Some other event besides ordinary conduction must have occurred since that date in order to produce the present state of things.

This is only one of the cases in which a consideration of the dissipation of energy leads to the determination of a superior limit to the antiquity of the observed order of things.

A very important class of problems is that in which there is a steady flow of heat into the body at one point of its surface, and out of it at another part. There is a certain distribution of temperature in all such cases, which if once established will not afterwards change: this is called the permanent distribution. If the original distribution differs from this, the effect of the diffusion of heat will be to cause the distribution of temperature to approximate without limit to this permanent distribution. Questions relating to the permanent distribution of temperature and the steady flow of heat are in general less difficult than those in which this state is not established.

Another important class of problems is that in which heat is supplied to a portion of the surface in a periodic manner, as in the case of the surface of the earth, which receives and emits heat according to the periods of day and night, and the longer periods of summer and winter.

The effect of such periodic changes of temperature at the surface is to produce waves of heat, which descend into the earth and gradually die away. The length of these waves is proportional to the square root of the periodic time. If we examine the wave at a depth such that the greatest heat occurs when it is coldest at the surface, then the extent of the variation of temperature at this depth is only $\frac{1}{23}$ of its value at the surface. In the rocks of this country this depth is about 25 feet for the annual variations.

In the diagram fig. 34 the distribution of temperature in the different strata is represented at two different times. If we suppose the figure to represent the diurnal variation of temperature, then the curves indicate the temperatures at

FIG. 34.

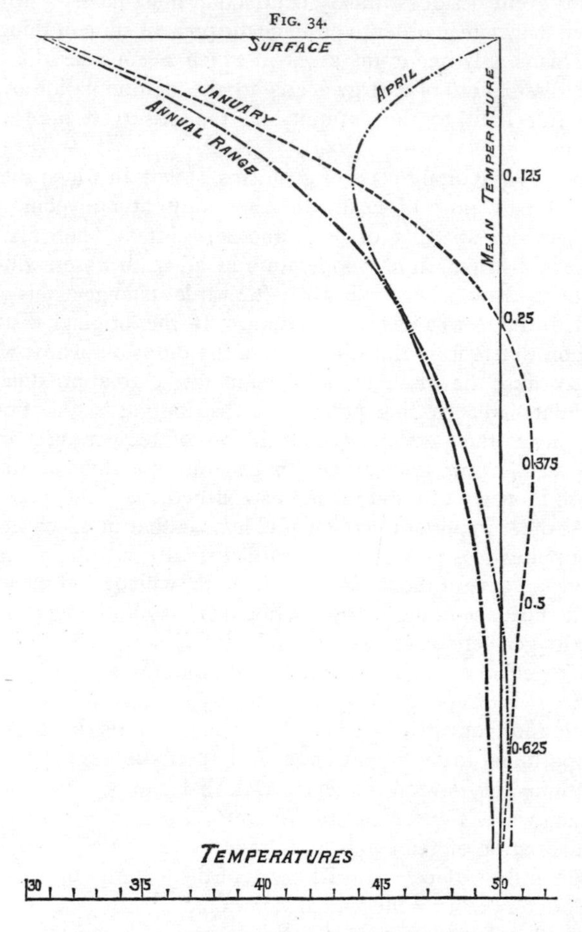

2 A.M. and 8 A.M. If we suppose it to represent the annual variation, then the curves correspond to January and April.

Since the depth of the wave varies as the square root of the periodic time, the wave-length of the annual variation of temperature will be about nineteen times the depth of those of the diurnal variation. At a depth of about 50 feet the variation of annual temperature is about a year in arrear.

The actual variation of temperature at the surface does not follow the law which gives a simple harmonic wave, but, however complicated the actual variation may be, Fourier shows how to decompose it into a number of harmonic waves of which it is the sum. As we descend into the earth these waves die away, the shortest most rapidly, so that we lose the irregularities of the diurnal variation in a few inches, and the diurnal variation itself in a few feet. The annual variation can be traced to a much greater depth; but at depths of 50 feet and upwards the temperature is sensibly constant throughout the year, the variation being less than the five-hundredth part of that at the surface.

But if we compare the mean temperatures at different depths, we find that as we descend the mean temperature rises, and that after we have passed through the upper strata, in which the periodic variations of temperature are observed, this increase of temperature goes on as we descend to the greatest depths known to man. In this country the rate of increase of temperature appears to be about 1° F. for 50 feet of descent.

The fact that the strata of the earth are hotter below than above shows that heat must be flowing through them from below upwards. The amount of heat which thus flows upwards in a year through a square foot of the surface can easily be found if we know the conductivity of the substance through which it passes. For several kinds of rock the conductivity has been ascertained by means of experiments made upon detached portions of the rock in the laboratory. But a still more satisfactory method, where it can be employed, is to make a register of the temperature at different depths throughout the year, and from this to determine the length of the annual wave of temperature, or its rate of decay. From either of these data the conductivity of the substance of the earth may be found without removing the rocks from their bed.

By observations of this kind made at different points of the earth's surface we might determine the quantity of heat which flows out of the earth in a year. This can be done only roughly at present, on account of the small number of places at which such observations have been made, but we know enough to be certain that a great quantity of heat escapes from the earth every year. It is not probable that any great proportion of this heat is generated by chemical action within the earth. We must therefore conclude that there is less heat in the earth now than in former periods of its existence, and that its internal parts were formerly very much hotter than they are now.

In this way Sir W. Thomson has calculated that, if no change has occurred in the order of things, it cannot have been more than 200,000,000 years since the earth was in the condition of a mass of molten matter, on which a solid crust was just beginning to form.

ON THE DETERMINATION OF THE THERMAL CONDUCTIVITY OF BODIES.

The most obvious method of determining the conductivity of a substance is to form it into a plate of uniform thickness, to bring one of its surfaces to a known temperature and the other to a known lower temperature, and to determine the quantity of heat which passes through the plate in a given time.

For instance, if we could bring one surface to the temperature of boiling water by a current of steam, and keep the other at the freezing temperature by means of ice, we might measure the heat transmitted either by the quantity of steam condensed, or by the quantity of ice melted.

The chief difficulty in this method is that the surface of the plate does not acquire the temperature of the steam or the ice with which it is in contact, and that it is difficult to ascertain its real temperature with the accuracy necessary for a determination of this kind.

Most of the actual determinations of conductivity have been made in a more indirect way—by observing the permanent distribution of temperature in a bar, one end of which is maintained at a high temperature, while the rest of its surface is exposed to the cooling effects of the atmosphere.

The temperatures of a series of points in the bar are ascertained by means of thermometers inserted into holes drilled in it, and brought into thermal connexion with its substance by means of fluid metal surrounding the bulbs.

In this way the rate of diminution of temperature with the distance can be ascertained at various points on the bar.

To determine the conductivity, we must compare the rate of variation of temperature with the flow of heat which is due to it. It is in the determination of this flow of heat that the indirectness of the method consists. The most trustworthy method of determining the flow of heat is that employed by Principal Forbes in his experiments on the conduction of heat in an iron bar.[1] He took a bar of exactly the same section and material as the experimental bar, and, after heating it uniformly, allowed it to cool in air of the same temperature as that surrounding the experimental bar. By observing the temperature of the cooling bar at frequent intervals of time, he ascertained the quantity of heat which escaped from the sides of the bar, this heat being measured in terms of the quantity of heat required to raise unit of volume *of the bar* one degree. This loss of heat depended of course on the temperature of the bar at the time, and a table was formed showing the loss from a linear foot of the bar in a minute at any temperature.

[1] *Trans. Roy. Soc. Edinb.* 1861–2.

Now, in the experimental bar the temperature of every part was known, and therefore the loss of heat from any given portion of the bar could be found by making use of the table. To determine the flow of heat across any particular section, it was necessary to sum up the loss of heat from all parts of the bar beyond this section, and when this was done, by comparing the flow of heat across the section with the rate of diminution of temperature per linear foot in the curve of temperature, the conductivity of the bar for the temperature of the section was ascertained. Principal Forbes found that the thermal conductivity of iron decreases as the temperature increases.

The conductivity thus determined is expressed in terms of the quantity of heat required to raise unit of volume *of the substance* one degree. If we wish to express it in the ordinary way in terms of the thermal unit as defined with reference to water at its maximum density, we must multiply our result by the specific heat of the substance, and by its density; for the quantity of heat required to raise unit of mass of the substance one degree is its specific heat, and the number of units of mass in unit of volume is the density of the substance.

As long as we are occupied with questions relating to the diffusion of heat and the waves of temperature in a single substance, the quantity on which the phenomena depend is the thermometric conductivity expressed in terms of the substance itself; but whenever we have to do with the effects of the flow of heat upon other bodies, as in the case of boiler plates, steam-condensers, &c., we must use a definite thermal unit, and express the calorimetric conductivity in terms of it. It has been shown by Professor Tyndall that the wave of temperature travels faster in bismuth than in iron, though the conductivity of bismuth is much less than that of iron. The reason is that the thermal capacity of the iron is much greater than that of an equal volume of bismuth.

Forbes was the first to remark that the order in which the metals follow one another in respect of thermal conductivity is nearly the same as their order as regards electric conductivity. This remark is an important one as regards certain metals, but it must not be pushed too far; for there are substances which are almost perfect insulators of electricity, whereas it is impossible to find a substance which will not transmit heat.

The electric conductivity of metals diminishes as the temperature rises. The thermal conductivity of iron also diminishes, but in a smaller ratio, as the temperature rises.

Professor Tait has given reasons for believing that the thermal conductivity of metals may be inversely proportional to their absolute temperature.

The electric conductivity of most non-metallic substances, and of all electrolytes and dielectrics, *increases* as the temperature rises. We have not sufficient data to determine whether this is the case as regards their thermal conductivity. According to the molecular theory of Chapter XXII. the thermal conductivity of gases increases as the temperature rises.

ON THE CONDUCTIVITY OF FLUIDS.

It is very difficult to determine the thermal conductivity of fluids, because the variation of temperature which is part of the phenomenon produces a variation of density, and unless the surfaces of equal temperature are horizontal, and the upper strata are the warmest, currents will be produced in the fluid which will entirely mask the phenomena of true conduction.

Another difficulty arises from the fact that most fluids have a very small conductivity compared with solid bodies. Hence the sides of the vessel containing the fluid are often the principal channel for the conduction of heat.

In the case of gaseous fluids the difficulty is increased by the greater mobility of their parts, and by the great variation of density with change of temperature. Their conductivity is extremely small, and the mass of the gas is generally small compared with that of the vessel in which it is contained. Besides this, the effect of direct radiation from the source of heat through the gas on the thermometer produces a heating effect which may, in some cases, completely mask the effect of true conduction. For all these reasons, the determination of the thermal conductivity of a gas is an investigation of extreme difficulty.

APPLICATIONS OF THE THEORY.

The great thermal conductivity of the metals, especially of copper, furnishes the means of producing many thermal effects in a convenient manner. For instance, in order to maintain a body at a high temperature by means of a source of heat at some distance from it, a thick rod of copper may be used to conduct the heat from the source to the body we wish to heat; and when it is desired to warm the air of a room by means of a hot pipe of small dimensions, the effect may be greatly increased by attaching copper plates to the pipe, which become hot by conduction, and expose a great heating surface to the air.

To ensure an exact equality of temperature in all the parts of a body, it may be placed in a closed chamber formed of thick sheet copper. If the temperature is not quite uniform outside this chamber, any difference of temperature between one part of the outer surface and another will produce such a flow of heat in the substance of the copper that the temperature of the inner surface will be very nearly uniform. To maintain the chamber at a uniform high temperature by means of a flame, as is sometimes necessary, it may be placed in a larger copper chamber, and so suspended by strings or supported on legs that very little heat can pass by direct conduction from the outer to the inner wall. Thus we have first an outer highly conducting shell of copper;

next a slowly conducting shell of air, which, however, tends to equalize the temperature by convection; then another highly conducting shell of copper; and lastly the inner chamber. The whole arrangement facilitates the flow of heat parallel to the walls of the chambers, and checks its flow perpendicular to the walls. Now differences of temperature within the chamber must arise from the passage of heat from without to within, or in the reverse direction, and the flow of heat along the successive envelopes tends only to equalize the temperature. Hence, by the arrangement of successive shells, alternately of highly conducting and slowly conducting matter, and still more if the slowly conducting matter is fluid, an almost complete uniformity of temperature may be maintained within the inner chamber, even when the outer chamber has all the heat applied to it at one point.

This arrangement was employed by M. Fizeau in his researches on the dilatation of bodies by heat.

CHAPTER XIX.

ON THE DIFFUSION OF FLUIDS.

THERE are many liquids which, when they are intermingled by being stirred together, remain mixed, and, though their densities are different, they do not separate from each other as oil and water do. When liquids which are capable of being permanently mixed are placed in contact with each other, the process of mixture goes on in a slow and gradual manner, and continues till the composition of the mixture is the same in every part.

Thus if we put a strong solution of any salt in the lower part of a tall glass jar, we may, by pouring water in a gentle stream on a small wooden float, fill up the jar with water without disturbing the solution. The process of diffusion will then go on between the water and the solution, and will continue for weeks or months, according to the nature of the salt and the height of the jar.

If the solution of the salt is strongly coloured, as in the case of sulphate of copper, bichromate of potash, &c., we may trace the process of diffusion by the gradual rise of the colour into the upper part of the jar, and the weakening of the colour in the lower part. A more exact method is that employed by Sir William Thomson, of placing a number of glass bubbles or beads in the jar, whose specific gravities are intermediate between that of the strong solution and that of water. At first the beads all float in the surface of separation between the two liquids, but as diffusion goes on they separate from each other, and indicate by their positions the specific gravity of the mixture at various depths. It is necessary to expel the air very thoroughly from both liquids by boiling before commencing this experiment. If this is not done, air separates from the liquids, and attaches itself in the form of small bubbles to the specific gravity beads, so that they no longer indicate the true specific gravity of the fluid in which they float. In order to determine the strength of the solution at any point, as indicated by one of the beads, we have only to measure the amount of the salt which must be added to a known quantity of pure water, in order to make the bead swim in the mixture.

Another method which has been suggested of observing the process of diffusion is to substitute for the jar a hollow glass prism having its refracting edge vertical, and to determine the refraction of a horizontal ray of light passing through the prism at different heights. To determine the strength of the solution indicated by an observed deviation of the ray of light, we have only to make a series of experiments with the same prism on solutions of known strength.

There are many pairs of liquids which do not diffuse into each other, and there are others in which the diffusion, after going on for some time, stops as soon as a certain small proportion of the heavier liquid has become mixed with the lighter, and a small proportion of the lighter has become mixed with the heavier.

In the case of gases, however, there is no such limitation. Every gas diffuses into every other gas, so that, however different the specific gravities of two gases may be, it is impossible to keep them from mixing if they are placed in the same vessel, even when the denser gas is placed below the rarer.

The fact of the diffusion of gases was first remarked by Priestley. The laws of the phenomena were first investigated by Graham. The rate at which the diffusion of any substance goes on is in every case proportional to the rate of variation of the strength of that substance in the fluid as we pass along the line in which the diffusion takes place. Each substance in the mixture flows from places where it exists in greater quantity to places where it is less abundant.

The law of diffusion of matter is therefore of exactly the same form as that of the diffusion of heat by conduction, and we can at once apply all that we know about the conduction of heat to assist us in understanding the phenomena of the diffusion of matter.

To fix our ideas, let us consider two very thin horizontal strata of the mixture, at a distance from each other equal to c, and we shall suppose that diffusion is going on at a constant rate through the part of the fluid between these strata.

Let the quantity of one of the substances, say the denser, in unit of volume of the mixture be Q in the lower stratum and q in the upper stratum, Q being greater than q; then diffusion of this substance will take place from the lower stratum to the upper through the intervening fluid, whose thickness is c. Let M be the quantity of the substance which passes upward in a time t through a horizontal area whose length is a and whose breadth is b; then we find, as in the case of heat,

$$\text{M} = \frac{k\,a\,b\,t}{c}(\text{Q}-q).$$

Here instead of H, a quantity of heat, we have M, a quantity of matter diffused; and instead of $T-S$, the difference of two temperatures, we have $Q-q$, the difference of the density of that kind of matter in two different strata. Finally, k, instead of meaning the thermal conductivity, means the coefficient of diffusion of the given substance in the given mixture.

If we determine k from this equation, we find

$$k = \frac{c}{a\,b\,t}\,\frac{M}{(Q-q)}.$$

Here $a\ b\ c$ are lines, Q and q are densities or quantities of matter in unit of volume, and M is a quantity of matter. Hence the fraction $\frac{M}{Q-q}$ is of the nature of a volume which is of three dimensions as regards length. We thus find that the dimensions of k, the coefficient of diffusion, are equal to the square of a length divided by a time.

Hence, in the experiment with the jar, the vertical distance between strata of corresponding densities, as indicated by the beads which float in them, varies as the square root of the time from the beginning of the diffusion.

When the mixture of two liquids or gases is effected in a more rapid manner by agitation or stirring, the only effect of the mechanical disturbance is to increase the area of the surfaces through which diffusion takes place. Instead of the surface of separation being a single horizontal plane, it becomes a surface of many convolutions, and of great extent, and in order to effect a complete mixture the diffusion has to extend only over the distance between the successive convolutions of this surface instead of over half the depth of the vessel.

Since the time required for diffusion varies as the square of the distance through which the diffusion takes place, it is easy to see that by stirring the solution in a jar along with the water above it, a complete mixture may be effected in a few seconds, which would have required months if the jar had been left undisturbed. That the mixture effected by stirring is not instantaneous may be easily seen by observing that during the operation the fluid appears to be full of streaks, which cause it to lose its transparency. This arises from the different indices of refraction of different portions of the mixture, which have been brought near each other by stirring. The surfaces of separation are so drawn out and convoluted that the whole mass has a woolly appearance, for no ray of light can pass without being turned many times out of its path.

The same appearance may also be observed when we mix hot water with cold, and even when very hot air is mixed with cold air. This shows that what is called the equalization of temperature by convection currents really takes place by conduction between portions of the substance brought near each other by the currents.

If we observe the process of diffusion with our most powerful microscopes, we cannot follow the motion of any individual portions of the fluids. We cannot point out one place in which the lower fluid is ascending, and another in which the upper fluid is descending. There are no currents visible to us, and the motion of the material substances goes on as imperceptibly as the conduction of heat or of electricity. Hence the motion which constitutes diffusion must be distinguished from those motions of fluids which we can trace by means of floating motes. It may be described as a motion of the fluids, not in mass, but by molecules.

We have not hitherto taken any notice of molecular theories, because we wish to draw a distinction between that part of our subject which depends only on the universal axioms of dynamics, combined with observations of the properties of bodies, and the part which endeavours to arrive at an explanation of these properties by attributing certain motions to minute portions of matter which are as yet invisible to us.

The description of diffusion as a molecular motion is one which we shall justify when we come to treat of molecular science. At present, however, we shall use the phrase 'molecular motion' as a convenient mode of describing the transference of a fluid when the motion of sensible portions of the fluid cannot be directly observed.

Graham observed that the diffusion both of liquids and gases takes place through porous solid bodies, such as plaster of Paris and pressed plumbago, at a rate not very much less than when no such body is interposed, and this even when the solid division is amply sufficient to check all ordinary currents, and even to support considerable differences of pressure on its opposite sides.

By taking advantage of the different velocities with which different liquids and gases pass through such substances, he was enabled to effect many important analyses and to arrive at new views of the constitution of various bodies.

But there is another class of cases in which a liquid or gas can pass through a diaphragm which is not in the ordinary sense porous. For instance, when carbonic acid gas is confined in a soap-bubble it gradually escapes. The liquid absorbs the gas at its inner surface, where it has the greatest density; and on the outside, where the density of the carbonic acid is less, the gas diffuses out into the atmosphere. During the passage of the gas through the film it is in the state of solution in water. It is also found that hydrogen and other gases can pass through a layer of caoutchouc. The ratios in which different gases pass through this substance are different from the ratios in which they percolate through porous plugs. Graham shows that the chemical relations between the gases and the caoutchouc determine these ratios, and that it is not through pores in the ordinary sense that the motion takes place.

According to Graham's theory, the caoutchouc is a colloïd substance—that is, one which is capable of being united, in a temporary and very loose manner, with various proportions of other substances, just as glue will form a jelly with various proportions of water. Another class of substances, which Graham calls crystalloïd, are distinguished from these by being always of definite composition, and not admitting of these temporary associations. When a colloïd substance has in differents parts of its mass different proportions of water, alcohol, or solutions of crystalloïd bodies, diffusion takes place through the colloïd substance, although no part of it can be shown to be in the liquid state.

On the other hand, a solution of a colloïd substance is almost incapable of diffusion through a porous solid, or through another colloïd substance. Thus, if a solution of gum in water containing salt be placed in contact with a solid jelly of gelatine containing alcohol, salt and water will be diffused into the gelatine, and alcohol will be diffused into the gum, but there will be no mixture of the gum and the gelatine.

There are certain metals whose relation to certain gases Graham explained by this theory. For instance, hydrogen can be made to pass through iron and palladium at a high temperature, and carbonic oxide can be made to pass through iron. The gases form colloïdal unions with the metals, and are diffused through them just as water is diffused through a jelly.

Graham made many determinations of the relative diffusibility of different salts. Accurate determinations of the coefficient of diffusion of liquids and gases are very much wanted, as they furnish important data for the molecular theory of these bodies. The most valuable determinations of this kind are those of the coefficient of diffusion between pairs of simple gases made by Professor J. Loschmidt of Vienna.[1]

He has determined the coefficient of diffusion in square

[1] Experimental-Untersuchungen über die Diffusion von Gasen ohne poröse Scheidewände. Sitzb. d. k. Akad. d. Wissensch. Bd. lxi. (March and July 1870.)

metres per hour for ten pairs of the most important gases. We shall consider these results when we come to the molecular theory of gases.

CHAPTER XX.

CAPILLARITY.

WE have hitherto considered the energy of a body as depending only on its temperature and its volume. The whole of the energy of gases, and the most important part of the energy of liquids, may be expressed in this way, but a very important part of the energy of a solid body may depend on the form which it is compelled to assume as well as on its volume. We shall return to this subject when treating of Elasticity and Viscosity, but we shall consider at present that part of the energy of a liquid which depends on the nature and extent of its surface.

To investigate the properties of a liquid surface, it is advisable to make use of the liquid in a form in which the surface is large in proportion to the volume. For this purpose, let us take common soapsuds, or Plateau's mixture of soap and glycerine, and blow a small bubble at the end of a tube with a bell mouth. A tobacco-pipe will answer, if the bore of the tube is large enough. After blowing the bubble at one end of the tube, place the other end near the flame of a candle. The bubble will contract and drive a current of air through the tube, as may be seen by its effect on the flame.

This shows that the bubble presses on the air within it, and is like an elastic bag. To enlarge the bubble by blowing into the tube, work must be done to force the air in, because the pressure inside the bubble is greater than that of the air outside. This work is stored up in the film of soapsuds, for it is able to blow the air out again with an equal force.

Hence a soap-bubble has a certain intrinsic energy depending in some way on its size.

The bubble consists of a thin stratum of liquid bounded by air on both sides. The intrinsic energy of the liquid, considered with reference to its volume and temperature only, will be the same for a bubble containing the same quantity of liquid, to whatever size it may be blown. But we know that work is done in making the bubble larger, and the only difference when the bubble is blown larger is that the surface of the liquid exposed to air is increased.

It appears from this, that in taking account of the energy of a system one of whose constituents is a liquid, we must take account, not only of the temperature and volume of the liquid, but of the extent and nature of its bounding surface.

In the case of the bubble the energy is greater the greater the extent of surface exposed to air. The amount of this energy for a soap-bubble at ordinary temperatures is, according to Plateau, about 5·6 gramme-metres per square metre in gravitation units. This is the amount of work required to blow a soap-bubble whose superficial extent is one square metre. As the soap-bubble has two surfaces exposed to air, the energy of a single surface is only 2·8 gramme-metres per square metre.

We shall call this the *superficial energy* of the soap-bubble. It is measured by the energy in unit of surface, and its dimensions when expressed in dynamical measure are therefore:

$$\frac{\text{energy}}{\text{area}} = \frac{L^2 M}{T^2} \frac{1}{L^2} = \frac{M}{T^2}$$

or it is of one dimension as regards mass, and of two dimensions inversely as regards time, and it is independent of the unit of length. Superficial energy depends on the nature of both the media of which the surface is a boundary. The media must be such as do not mix with each other, otherwise diffusion occurs, and the surface of separation becomes indefinite; but there is a coefficient of superficial energy for every surface which separates two liquids which do not mix—a liquid and a gas, or its own vapour; and for the surface which separates a liquid and a solid, whether it dissolves the solid or not. There is also a coefficient of superficial energy for the surface separating a gas and a solid, or two solids; but as any two gases diffuse into each other, they can have no surface of separation.

Superficial Tension.

When the area of the surface is increased in any way, work must be done; and when the surface is allowed to contract, it does work on other bodies. Hence it acts like a stretched sheet of india-rubber, and exerts a tension of the same kind. The only difference is, that the tension in the sheet of india-rubber depends on the amount of stretching, and may be greater in one direction than in a direction at right angles to it, whereas the tension in the soap-bubble remains the same however much the film is extended, and the tension at any point is the same in all directions.

If we draw a straight line, P Q, across the surface A B D C, and if the whole tension exerted by the surface across the line P Q is F, then the superficial tension is measured by the tension across unit of length of the line P Q; or, since F is the tension across the whole line, if T is the superficial tension across unit of length,

FIG. 35.

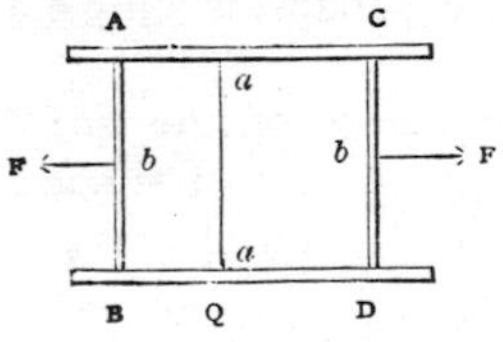

$$F = T \cdot PQ.$$

Now let us suppose that the lines A B and C D were originally in contact, and that the surface A B D C was produced by drawing C D away from A B by the action of the force F.

If we suppose A B and B C to be rods wet with soapsuds, placed between two parallel rods A C and B D and then drawn asunder, the soap film A B D C will be formed. If S is the superficial energy of the film per unit of area, then the work done in drawing it out will be S . A B . A C. But if F is the force required to draw A B from C D, the same work may be written F . A C, or, putting for F its value in terms of T, and equating the two expressions for the work,

$$S . AB . AC = T . PQ . AC$$

$$\text{or} \quad = T . AB . AC.$$

Hence

$$S = T,$$

or the numerical value of the superficial energy per unit of area is equal to that of the superficial tension per unit of length. This quantity is usually called the Coefficient of Capillarity, because it was first considered with reference to the ascent of liquids in capillary tubes. These tubes derived their name from the smallness of their bore, which would only admit a hair (*capilla*). I have used the phrases 'superficial energy' and 'superficial tension' because I think they help us to direct our attention to the facts, and to understand the various phenomena of liquid surfaces better than a name which is purely technical, and which has already done a great deal of harm when used without being understood. If by the help of this treatise, or otherwise, anyone has obtained a clear conception of the real phenomena called Capillary Attraction and Capillarity, he may use these words quite freely. The theory as we shall state it does not differ essentially from that originally given by Laplace, though by the free use of the idea of superficial tension we avoid some of the mathematical operations which are required to deduce the phenomena from the hypothesis of molecular attractions.

We shall now suppose that the superficial tension is known for the surfaces which bound every pair of the media with which we have to do. For instance, we may denote by T_{ab} the superficial tension of the surface which separates the medium a from the medium b.

Let there be three fluid media, *a*, *b*, *c*, and let the surface of separation between *a* and *b* meet the surface of separation between *b* and *c* along a line of any form having continuous curvature. Let O be a point in this line, and let the plane of the paper represent a section perpendicular to the line.

The three tensions T_{ab}, T_{bc}, and T_{ca} must be in equilibrium along this line, and, since we know these tensions, we can easily determine the angles which they make with each other. In fact, if we construct a triangle A B C having lines proportional to these tensions for its sides, the exterior angles of this triangle will be equal to the angles formed by the three surfaces of separation which meet in a line.

By trigonometry, if A B C are the angles of the edges formed by the media *a b c*, then

$$\frac{T_{bc}}{\sin A} = \frac{T_{ca}}{\sin B} = \frac{T_{ab}}{\sin C}.$$

It appears from this that whenever three fluid media are

Fig. 36.

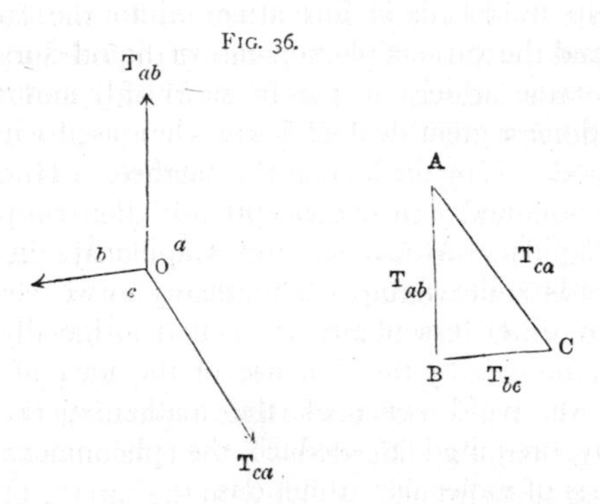

in contact and in equilibrium, the angles between their surfaces of separation depend only on the values of the superficial tensions of these three surfaces, and are therefore always the same for the same three fluids.

But it is not always possible to construct a triangle with

three given lines as its sides. If any one of the lines is greater than the sum of the other two, the triangle cannot be formed. For the same reason, if any one of the three superficial tensions is greater than the sum of the other two, the three fluids cannot be in equilibrium in contact.

For instance, if the tension of the surface separating air and water is greater than the sum of the tensions of the surfaces separating air and oil, and oil and water, then a drop of oil cannot be in equilibrium on the surface of water. The edge of the drop, where the oil meets the air and the water, becomes thinner and thinner; but even when the angle is reduced to the thinnest edge, the tension of the free surface of the water exceeds the tensions of the two surfaces of the oil, so that the oil is drawn out thinner and thinner, till it covers a vast expanse of water. In fact, the process may go on till the oil becomes so thin, and contains so small a number of molecules in its thickness, that it no longer has the properties of the liquid in mass.

When a solid body is in contact with two fluids, then if the tension of the surface separating the solid from the first fluid exceeds the sum of the tensions of the other two surfaces, the first fluid will gather itself up into a drop, and the second will spread over the surface. If one of the fluids is air, and the other a liquid, then the liquid, if it corresponds to the first fluid mentioned above, will stand in drops without wetting the surface; but if it corresponds to the second, it will spread itself over the whole surface, and wet the solid.

When the tension of the surface separating the two fluids is greater than the difference of the tensions of the surfaces separating them from the solid, then the surface of separation of the two fluids will be inclined at a finite angle to the surface of the solid. Thus, if a and b are the two fluids, and c the solid, then to find the angle of contact P O Q we must make $PO = T_{ab}$, and $OQ = T_{bc} - T_{ac}$. This angle is called the angle of capillarity.

ON THE RISE OF A LIQUID IN A TUBE.

Let a be a liquid in a tube of a substance c, whose radius is r. Let the fluid b be air or any other fluid. Let α be the angle of capillarity. The circumference of the tube is $2\ \pi\ r$. All round this circumference there is a tension T_{ab} acting at an angle inclined α to the vertical, and therefore the whole vertical force is

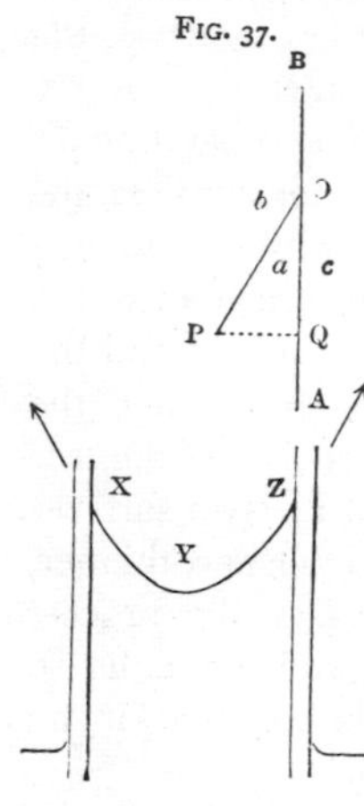

FIG. 37.

$$2\ \pi\ r\ T_{ab} \cos \alpha.$$

If this force raises the liquid to a height h, then, neglecting the weight of the sides of the hollow portion X Y Z, the weight of fluid supported is

$$\pi\ \rho\ g\ r^2\ h.$$

Equating this force to the weight which it supports, we find

$$h = 2\ \frac{T_{ab} \cos \alpha}{\rho\ g\ r}.$$

Hence the height to which the fluid is drawn up is inversely as the radius of the tube.

A liquid is drawn up in the same way in the space between two parallel plates separated by a distance d. If we now suppose fig. 38 to represent a section of the film of liquid, the horizontal breadth of which is l, then the surface-tension of the liquid on the line which bounds the wet and dry parts of each plate is $T\ l$, and this force acts at an angle α with the vertical. The whole force, therefore, arising from the surface-tension, and tending to raise the liquid, is

$$2\ T\ l \cos \alpha.$$

The weight of the liquid raised is

$$\rho\ g\ h\ l\ d.$$

Equating the force to the weight which it supports, we find

$$h = 2 \frac{T \cos \alpha}{\rho g d}.$$

This expression differs from that for the height in a cylindrical tube only by the substitution of d, the distance between the parallel plates, for r, the radius of the tube. Hence the height to which a liquid will ascend between two plates is equal to the height to which it rises in a tube whose radius is equal to the distance between the plates, or whose diameter is twice that distance.

A remarkable application of the principles of thermodynamics to capillary phenomena has recently been made by Sir W. Thomson.[1] Let a fine tube be placed in a liquid, and let the whole be placed in a vessel from which air is exhausted, so that the whole space above the liquid becomes filled with its vapour and nothing else.

Fig. 38.

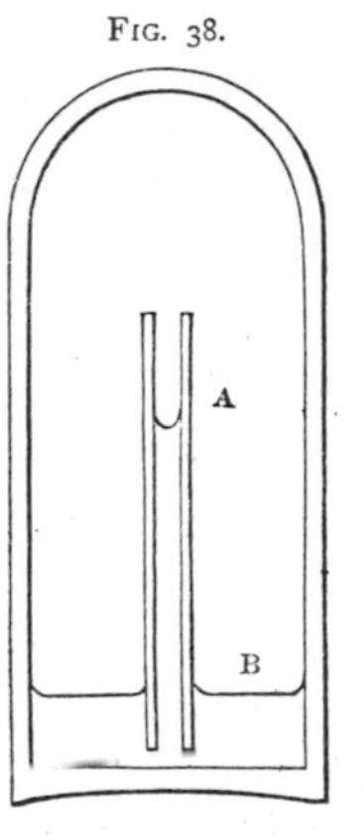

Let the permanent level of the liquid be at A in the small tube, and at B in the vessel, and let us suppose the temperature the same throughout the apparatus.

There is a state of equilibrium between the liquid and its vapour, both at A and at B; otherwise evaporation or condensation would occur, and the permanent state would not exist.

Now the pressure of the vapour at B exceeds that at A by the pressure due to a column of the vapour of the height A B.

It follows that the vapour is in equilibrium with the liquid at a lower pressure where the surface of the liquid is concave, as at A, than where it is plane, as at B.

Now let the lower end of the tube be closed, and let some of the liquid be taken out of it, so that the liquid in the tube does not reach up to the point A.

[1] *Proceedings of the Royal Society of Edinburgh*, Feb. 7, 1870.

Then vapour will condense inside the tube, owing to the concavity of its surface, and this will go on till it is filled with liquid up to the level A, the same as if it had been open at the bottom.

Hence, if at any point of a concave liquid surface r and are the principal radii of curvature of the surface, and if the pressure of vapour in equilibrium with a plane surface of its liquid at the given temperature is ϖ, and if p is the pressure of equilibrium of the vapour in contact with the curved surface,

$$p = \varpi - \frac{T\sigma}{\rho-\sigma}\left(\frac{1}{r}+\frac{1}{r'}\right),$$

where σ is the density of the vapour, and ρ that of the liquid.

If h is the height to which the liquid would rise in virtue of the curvature of its surface in a capillary tube, and if $\mathfrak{H}$ is the height of a homogeneous atmosphere of the vapour,

$$p = \varpi\left(1-\frac{h}{\mathfrak{H}}\right).$$

Sir W. Thomson has calculated that in a tube whose radius is about a thousandth of a millimetre, and in which water would rise about thirteen metres above the plane level, the equilibrium pressure of aqueous vapour would be less than that on a plane surface of water by about a thousandth of its own amount.

He thinks it probable that the moisture which vegetable substances, such as cotton, cloth, &c., acquire from air at temperatures far above the dew point may be explained by the condensation of water in the narrow tubes and cells of the vegetable structure.

In the case of a spherical bubble of steam in water, the increase or diminution of the diameter depends on the temperature and pressure of the vapour within; and the condition that ebullition may take place is that the pressure of saturated vapour at the temperature of the liquid must exceed the actual pressure of the liquid by a pressure equal to that of a column of the liquid of the height to which it would ascend in a tube whose section is equal to that of the bubbles.

If the liquid contains any gas in solution, or any liquid more volatile than itself, or if air or steam is made to bubble up through the liquid, then bubbles will be formed of a visible diameter, and the ebullition will be kept up by evaporation at the surface of these bubbles. But if, by long boiling or otherwise, the liquid is deprived of any substance more volatile than itself, and if the sides of the vessel in which it is contained are such that the liquid adheres closely to them, so that bubbles, if formed at the surface of the vessel, will rather collect into a spherical form than spread along the surface, then the temperature of the liquid may be raised far above the boiling point, and when boiling at last occurs, it goes on in an almost explosive manner, and the liquid 'bumps' violently on the bottom of the vessel.

The highest temperature to which water may be raised under the atmospheric pressure without ebullition cannot be said to be accurately known, for every improvement in the arrangements for getting rid of condensed air, &c., has made it possible to raise liquid water to a higher temperature. In an experiment due to Dufour, the water, instead of being allowed to touch the sides of the vessel, is dropped into a mixture of linseed oil and oil of cloves, which has nearly the same density as itself. By this means, drops of liquid water may sometimes be observed swimming in the mixture at a temperature of 356° F. The pressure of aqueous vapour is at this temperature nearly ten atmospheres, or about 147 pounds' weight on the square inch. Hence the cohesion of the water must be able to support at least 132 pounds weight on the square inch.

We may also apply Sir W. Thomson's principle to the case of evaporation from a small drop. In this case the surface of the liquid is convex, so that if r is the radius of the drop,

$$p = \varpi + \mathrm{T}\frac{\tau}{\rho - \sigma} \cdot \frac{2}{r}.$$

Here ϖ is the pressure of saturated vapour corresponding to the temperature when the surface of the liquid is plane, and p is the pressure of vapour required to prevent the drop from evaporating. A small drop will therefore evaporate in air containing so much moisture that condensation would take place on a flat surface.

Hence, if a vapour free from suspended particles, and not in contact with any solid body except such as are warmer than itself, is cooled by expansion, it is probable that the suggestion of Prof. J. Thomson at p. 126 might be verified, and that the vapour might be cooled below its ordinary point of condensation without liquefaction, for the first effect of condensation would be to produce excessively small drops, and these, as we have seen, would not tend to increase unless the vapour surrounding them were more than saturated.

The formation of cloud in vapour often appears very sudden, as if it had been at first retarded by some cause of this kind, so that when at last the cloud is formed condensation occurs with great rapidity, reminding us of the converse phenomenon of the rapid boiling of an overheated liquid.

The drops in a cloud, for the same reason, cannot remain of the same size, even if they are not jostled against each other, for the smaller drops will evaporate, while the larger ones are increased by condensation, so that visible drops will be formed by pure condensation without any necessity for the coalescence of smaller drops.

Up to this point we have not considered the effect of heat on the superficial tension of liquids. In all liquids on which experiments have been made the superficial tension diminishes as the temperature rises, being greatest at the freezing point of the substance, and vanishing altogether at the critical point where the liquid and gaseous states become continuous.

It appears, therefore, that the phenomenon is intimately related to the apparent discontinuity of the liquid and gaseous states, and that it must be studied in connexion with the conditions of evaporation and the phenomenon called latent heat. Much light will probably be thrown on all these subjects by investigations which as yet can hardly be said to be begun.

Sir W. Thomson has applied the principles of thermodynamics to the case of a film of water extended by a force applied to it, and has shown that in order to maintain the temperature of the film constant an amount of heat must be supplied to it nearly equal in dynamical measure to half the work done in stretching the film.

In fact, the third thermodynamical relation (p. 167) may be applied at once to the case by making the following substitutions: for 'pressure' put 'superficial tension,' and for 'volume' put 'area.'

We thus find that the latent heat of extension of unit of area is equal to the product of the absolute temperature and the decrement of superficial tension per degree of temperature. At ordinary temperatures it appears from experiment that this product is about half the superficial tension. Hence the latent heat of extension in dynamical measure is about half the work spent in producing the extension.

The student may also adapt the investigation of latent heat as given at p. 172 to the case of the extension of a liquid film.

The following table, taken from the memoir of M. Quincke, gives the superficial tension of different liquids in contact with air, water, and mercury. The tension is measured in grammes weight per linear metre, and the temperature is 20° C.

Table of Superficial Tension at 20° C.

Liquid	Sp. gravity	Tension of surface separating the liquid from		
		Air	Water	Mercury
Water	1·0	8·253	0	42·58
Mercury	13·5432	55·03	42·58	0
Bisulphide of Carbon . .	1·2687	3·274	4·256	37·97
Chloroform	1·4878	3·120	3·010	40·71
Alcohol	0·7906	2·599	—	40·71
Olive Oil	0·9136	3·760	2·096	34·19
Turpentine	0·8867	3·030	1·177	25·54
Petroleum	0·7977	3·233	2·834	28·94
Hydrochloric Acid . .	1·1	7·15	—	38·41
Solution of Hyposulphite of Soda	1·1248	7·903	—	45·11

It appears from this table that water has the greatest superficial tension of all ordinary liquids. For this reason it is very difficult to preserve a surface of pure water. It is sufficient to touch any part of the surface of pure water with a greased rod to reduce its tension considerably. The smallest quantity of any kind of oil immediately spreads itself over the surface, and completely alters the superficial tension. Hence the importance in all experiments on superficial tension of having the vessel thoroughly clean. This has been well pointed out by Mr. Tomlinson in his researches on the 'cohesion figures of liquids.'

When one of the liquids is soluble in the other, the effects of superficial tension are very remarkable. For instance, if a drop of alcohol be placed on the surface of a thin layer of water, the tension is immediately reduced to 2·6, where the alcohol is pure, and varies from this value to 8·25, where the water is pure. The result is that the equilibrium of the surface is destroyed, and the superficial film of the liquid is set in motion from the alcohol towards the water, and if the water is shallow this motion of the surface will drag the whole of the water with it, so as to lay bare part of the bottom of the vessel. A dimple may be formed on the surface of water by bringing a drop of ether close to the surface. The vapour of the ether condensed on the surface of the water is sufficient to cause the outward current mentioned above.

Wine contains alcohol and water, and when it is exposed to the air the alcohol evaporates faster than the water, so that the superficial layer becomes weaker. When the wine is in a deep vessel, the strength is rapidly equalized by diffusion; but in the case of the thin layer of wine which adheres to the sides of a wineglass, the liquid rapidly becomes more watery. This increases the superficial tension at the sides of the glass, and causes the surface to be dragged from the strong wine to the weak. The watery portion is always uppermost, and creeps up the sides of the glass, dragging the stronger wine after it till the quantity of the fluid becomes so great that the different portions mix, and the drop runs down the side.

This phenomenon, known as the tears of strong wine, was first explained on these principles by Professor James Thomson. It is probable that it is referred to in Proverbs xxiii. 31, as an indication of the strength of the wine. The motion ceases in a stoppered bottle as soon as enough of vapour of alcohol has been formed in the bottle to be in equilibrium with the liquid alcohol in the wine.

The fatty oils have a greater superficial tension than turpentine, benzol, or ether. Hence if there is a greasy spot on a piece of cloth, and if one side of it is wetted with one of these substances, the tension is greatest on the side of the grease, and the portions consisting of mixtures of benzol and grease move from the benzol towards the grease.

If in order to cleanse the grease-spot we begin by wetting the middle of the spot with benzol, we drive away the grease into the clean part of the cloth. The benzol should therefore be applied first in a ring all round the spot, and gradually brought nearer to the centre of the spot, and a fibrous substance, such as blotting-paper, should be placed in contact with the cloth, so that when the grease is chased by the benzol to the middle of the spot it may make its escape into the blotting-paper, instead of remaining in globules on the surface, ready to return into the cloth when the benzol evaporates.

Another very effectual method of getting rid of grease-spots is founded on the fact that the superficial tension of a substance always diminishes as the temperature rises. If, therefore, the temperature is different at different parts of a greasy cloth, the grease tends to move from the hot parts to the cold. We therefore apply a hot iron to one side of the cloth, and blotting-paper to the other, and the grease is driven into the blotting-paper. If there is blotting-paper on both sides it will be found that the grease is driven mainly into that on the opposite side from the hot iron.

CHAPTER XXI.

ON ELASTICITY AND VISCOSITY.

On Stresses and Strains.

When the form of a connected system is altered in any way, the alteration of form is called a Strain. The force or system of forces by which this strain is produced or maintained is called the Stress corresponding to the strain. There are different kinds of strains, and different kinds of stresses corresponding to them.

The only case which we have hitherto considered is that in which the three longitudinal stresses are equal. This kind of stress is called Hydrostatic Pressure, and is the only kind which can exist in a fluid at rest. The pressure is the same in whatever direction it is estimated.

A very important kind of stress is called Shearing Stress: it is compounded of two equal longitudinal stresses, one being a tension and the other a pressure acting at right angles to each other. When a pair of scissors is employed to cut anything, the two blades produce a shearing stress in the material between them, tending to make one portion slide over the other.

FIG. 39.

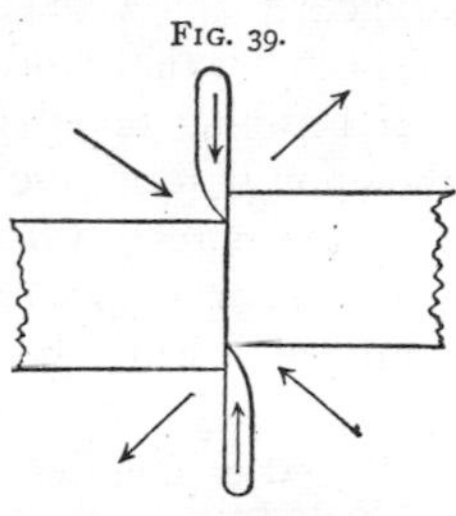

We have now to consider the properties of bodies when acted on by this kind of stress.

A body which when subjected to a stress experiences no strain would, if it existed, be called a Perfectly Rigid Body. There are no such bodies, and this definition is given only to indicate what is meant by perfect rigidity.

A body which when subjected to a given stress at a given temperature experiences a strain of definite amount, which does not increase when the stress is prolongèd, and which disappears completely when the stress is removed, is called a Perfectly Elastic Body.

Gases and liquids, and perhaps most solids, are perfectly elastic as regards stress uniform in all directions, but no substance which has yet been tried is perfectly elastic as regards shearing stress, except perhaps for exceedingly small values of the stress.

Now suppose that stresses of the same kind, but of continually increasing magnitude, are applied to a body in succession. As long as the body returns to its original form when the stress is removed it is said to be perfectly elastic.

If the form of the body is found to be permanently altered when the stress exceeds a certain value, the body is said to be soft, or plastic, and the state of the body when the alteration is just going to take place is called the Limit of Perfect Elasticity.

If the stress be increased till the body breaks or gives way altogether, the value of the stress is called the Strength of the body for that kind of stress.

If breaking takes place before there is any permanent alteration of form, the body is said to be Brittle.

If the stress, when it is maintained constant, causes a strain or displacement in the body which increases continually with the time, the substance is said to be Viscous.

When this continuous alteration of form is only produced by stresses exceeding a certain value, the substance is called a solid, however soft it may be. When the very smallest stress, if continued long enough, will cause a constantly increasing change of form, the body must be regarded as a viscous fluid, however hard it may be.

Thus, a tallow candle is much softer than a stick of sealing-wax; but if the candle and the stick of sealing-wax are laid horizontally between two supports, the sealing-wax will in a few weeks in summer bend with its own weight, while the candle remains straight. The candle is therefore a soft solid, and the sealing-wax a very viscous fluid.

What is required to alter the form of a soft solid is a sufficient force, and this, when applied, produces its effect at once. In the case of a viscous fluid it is *time* which is required, and if enough time is given, the very smallest force will produce a sensible effect, such as would require a very large force if suddenly applied.

Thus a block of pitch may be so hard that you cannot make a dint in it by striking it with your knuckles; and yet it will, in the course of time, flatten itself out by its own weight, and glide down hill like a stream of water.

A glass fibre was found by M. F. Kohlrausch [1] to become more and more twisted when constantly acted on by

[1] *Pogg.* 1863.

the small twisting force arising from the action of the earth on a little magnet suspended by the fibre. I have found slow changes in the torsion of a steel wire going on for many days after it had received a slight permanent twist, and Sir W. Thomson[1] has investigated the viscosity of other metals.

There are instances of viscosity among very hard bodies. Returning to our former example, pitch : we may mix it in various proportions with tar so as to form a continuous series of compounds passing from the apparently solid condition of pitch to the apparently fluid condition of tar, which may be taken as a type of a viscous fluid. By mixing the tar with turpentine the viscosity may be still further reduced, and so we may form a series of fluids of diminishing viscosity till we arrive at the most mobile fluids, such as ether.

DEFINITION OF THE COEFFICIENT OF VISCOSITY.

Consider a stratum of the substance of thickness c, contained between the horizontal fixed plane A B and the plane C D, which is moving horizontally from C towards D, with the velocity V. Let us suppose that the substance between the two planes is also in motion, the stratum in contact with C D moving with velocity V, while the velocity of any intermediate stratum is proportional to its height above A B.

FIG. 40.

C D

A B

The substance between the planes is undergoing shearing strain, and the rate at which this strain is increasing is measured by the velocity V of the upper plane, divided by the distance c between the planes, or $\frac{V}{c}$.

The stress F is a shearing stress, and is measured by the horizontal force exerted by the substance on unit of area

[1] *Proc. Roy. Soc.* May 18, 1865.

of either of the planes, and acting from A to B on the lower plane, and from D to C on the upper.

The ratio of this force to the rate of increase of the shearing stress is called the coefficient of viscosity, and is denoted by the symbol μ. We may therefore write

$$F = \mu \frac{V}{c}.$$

If R is the amount of this force on a rectangular area of length a and breadth b,

$$R = a\, b\, F$$

$$= \mu \frac{a\, b}{c} V$$

and

$$\mu = \frac{R\, c}{V\, a\, b}.$$

If V, a, b, and c are each unity, then $\mu = R$.

Definition.—The viscosity of a substance is measured by the tangential force on the unit of area of either of two horizontal planes at the unit of distance apart, one of which is fixed, while the other moves with the unit of velocity, the space between being filled with the viscous substance.

The dimensions of μ may be easily determined. If R is the moving force which would generate a certain velocity v in the mass M in the time t, then $R = \frac{M\, v}{t}$, and

$$\mu = \frac{M\, v\, c}{t\, V\, a\, b}.$$

Here a, b, c are lines, and V and v are velocities, so that the dimensions of μ are $[M\, L^{-1}\, T^{-1}]$, where M, L, and T are the units of mass, length, and time.

When we wish to express the absolute forces called into play by the viscosity of a substance, we must use the ordinary unit of mass (a pound, a grain, or a gramme); but if we wish only to investigate the motion of the viscous substance, it is convenient to take as our unit of mass that of unit of volume of the substance itself. If ρ is the density of the substance, or the mass of unit of volume, the viscosity ν measured in this kinematic way is related to μ, its value by the former, or dynamical method, by the equation

$$\mu = \nu \rho.$$

The dimensions of ν, the kinematic viscosity, are $[L^2 T^{-1}]$.

Investigations of the value of viscosity have been made, for solids by Sir W. Thomson; for liquids by Poiseuille, Graham, O. E. Meyer, and Helmholtz; and for gases by Graham, Stokes, O. E. Meyer, and myself.

I find the value of μ for air at $\theta°$ Centigrade to be

$$\mu = \cdot 0001878\ (1 + \cdot 00366\theta),$$

the centimetre, gramme, and second being units.

In British measure, using the foot, the grain, and the second, and Fahrenheit's thermometer, this becomes

$$\mu = \cdot 000179\ (461 + \theta).$$

The viscosity μ is proportional to the absolute temperature, and independent of the pressure, being the same for a pressure of half an inch as for a pressure of thirty inches of mercury. The significance of this remarkable result will be seen when we come to the molecular theory of gases.

The kinematic measure, ν, of the viscosity is found by dividing μ by the density. It is therefore directly proportional to the square of the absolute temperature, and inversely proportional to the pressure.

The value of μ for hydrogen is less than half that for air. Oxygen, on the other hand, has a viscosity greater than that of air. That of carbonic acid is less than that of air.

It appears, from the calculations of Professor Stokes, combined with the value of the viscosity of air given above, that a drop of water falling through air one thousand times rarer than itself (which we may suppose to be the case at the ordinary height of a cloud) would fall about $\frac{8}{10}$ of an inch in a second if its diameter were the thousandth part of an inch. If the diameter of the drop were only one ten-thousandth of an inch the rate at which it would make its way through the air would be a hundred times smaller, or half an inch in a minute. If a cloud is formed of little drops of water of this size, their motion through the air would be so slow that it would escape observation, and the motion of the cloud, so far as it can be observed, would be the same as that of the air in that place. In fact, the settling down through the air of any very small particles, such as the fine spray of waves or waterfalls, and all kinds of dust and smoke, is a very slow process, and the time of settling down through a given distance varies inversely as the square of the dimensions of the particles, their density and figure being the same. If, however, a cloud of fine dust contains so many particles that the mass of a cubic foot of dusty air is sensibly greater than that of a cubic foot of pure air, the dusty air will descend in mass below the level of the pure air like a fluid of greater density, so that a room may have its lower half filled with dusty air separated by a level surface from the pure air above.

There are some kinds of fogs the mean density of which is greater than that of the purer air in the neighbourhood, and these lie like lakes in hollows, and pour down valleys like streams. On the other hand, the mean density of a cloud may be less than that of the surrounding air, and it will then ascend.

In the case of smoke, both the air and the sooty particles are heated by the fire before they escape into the atmosphere, but, independently of this kind of heating, if the sun shines on a cloud of dust or smoke, the particles absorb heat, which they communicate to the air round them, and thus, though the particles themselves remain much denser than the air in the neighbourhood, they may cause the cloud which they form to appropriate so much of the sun's heat that it becomes lighter as a whole than the surrounding pure air, and so rises.

In the case of a cloud of watery particles, besides this

kind of action, there is another, depending on the evaporation from the surface of the little drops. The vapour of water is much rarer than air, and damp air is lighter than dry air at the same temperature and pressure. Hence the little drops make the air of the cloud damp, and if the mean density of the cloud is by this means made less than that of the surrounding air, the cloud will ascend.

CHAPTER XXII.

ON THE MOLECULAR THEORY OF THE CONSTITUTION OF BODIES.

WE have already shown that heat is a form of energy—that when a body is hot it possesses a store of energy, part at least of which can afterwards be exhibited in the form of visible work.

Now energy is known to us in two forms. One of these is Kinetic Energy, the energy of motion. A body in motion has kinetic energy, which it must communicate to some other body during the process of bringing it to rest. This is the fundamental form of energy. When we have acquired the notion of matter in motion, and know what is meant by the energy of that motion, we are unable to conceive that any possible addition to our knowledge could explain the energy of motion, or give us a more perfect knowledge of it than we have already.

There is another form of energy which a body may have, which depends, not on its own state, but on its position with respect to other bodies. This is called Potential Energy. The leaden weight of a clock, when it is wound up, has potential energy, which it loses as it descends. It is spent in driving the clock. This energy depends, not on the piece of lead considered in itself, but on the position of the lead with respect to another body—the earth—which attracts it.

In a watch, the mainspring, when wound up, has potential energy, which it spends in driving the wheels of the watch. This energy arises from the coiling up of the spring, which alters the relative position of its parts. In both cases, until the clock or watch is set agoing, the existence of potential energy, whether in the clock-weight or in the watch-spring, is not accompanied with any visible motion. We must therefore admit that potential energy can exist in a body or system all whose parts are at rest.

It is to be observed, however, that the progress of science is continually opening up new views of the forms and relations of different kinds of potential energy, and that men of science, so far from feeling that their knowledge of potential energy is perfect in kind, and incapable of essential change, are always endeavouring to explain the different forms of potential energy; and if these explanations are in any case condemned, it is because they fail to give a sufficient reason for the fact, and not because the fact requires no explanation.

We have now to determine to which of these forms of energy heat, as it exists in hot bodies, is to be referred. Is a hot body, like a coiled-up watch-spring, devoid of motion at present, but capable of exciting motion under proper conditions? or is it like a fly-wheel, which derives all its tremendous power from the visible motion with which it is animated?

It is manifest that a body may be hot without any motion being visible, either of the body as a whole, or of its parts relatively to each other. If, therefore, the body is hot in virtue of motion, the motion must be carried on by parts of the body too minute to be seen separately, and within limits so narrow that we cannot detect the absence of any part from its original place.

The evidence for a state of motion, the velocity of which

must far surpass that of a railway train, existing in bodies which we can place under the strongest microscope, and in which we can detect nothing but the most perfect repose, must be of a very cogent nature before we can admit that heat is essentially motion.

Let us therefore consider the alternative hypothesis—that the energy of a hot body is potential energy, or, in other words, that the hot body is in a state of rest, but that this state of rest depends on the antagonism of forces which are in equilibrium as long as all surrounding bodies are of the same temperature, but which as soon as this equilibrium is destroyed are capable of setting bodies in motion. With respect to a theory of this kind, it is to be observed that potential energy depends essentially on the relative position of the parts of the system in which it exists, and that potential energy cannot be transformed in any way without some change of the relative position of these parts. In every transformation of potential energy, therefore, motion of some kind is involved.

Now we know that whenever one body of a system is hotter than another, heat is transferred from the hotter to the colder body, either by conduction or by radiation. Let us suppose that the transfer takes place by radiation. Whatever theory we adopt about the kind of motion which constitutes radiation, it is manifest that radiation consists of motion of some kind, either the projection of the particles of a substance called caloric across the intervening space, or a wave-like motion propagated through a medium filling that space. In either case, during the interval between the time when the heat leaves the hot body and the time when it reaches the cold body, its energy exists in the intervening space in the form of the motion of matter.

Hence, whether we consider the radiation of heat as effected by the projection of material caloric, or by the undulations of an intervening medium, the outer surface of a hot body must be in a state of motion, provided any cold body is in its neighbourhood to receive the radiations which it emits. But we have no reason to believe that the presence of a cold body is essential to the radiation of heat by a hot one. Whatever be the mode in which the hot body shoots forth its heat, it must depend on the state of the hot body alone, and not on the existence of a cold body at a distance, so that even if all the bodies in a closed region were equally hot, every one of them would be radiating heat; and the reason why each body remains of the same temperature is, that it receives from the other bodies exactly as much heat as it emits. This, in fact, is the foundation of Prevost's Theory of Exchanges. We must therefore admit that at every part of the surface of a hot body there is a radiation of heat, and therefore a state of motion of the superficial parts of the body. Now this motion is certainly invisible to us by any direct mode of observation, and therefore the mere fact of a body appearing to be at rest cannot be taken as a demonstration that its parts may not be in a state of motion.

Hence part, at least, of the energy of a hot body must be energy arising from the motion of its parts, or kinetic energy.

The conclusion at which we shall arrive, that a very considerable part of the energy of a hot body is in the form of motion, will become more evident when we consider the thermal energy of gases.

Every hot body, therefore, is in motion. We have next to enquire into the nature of this motion. It is evidently not a motion of the whole body in one direction, for however small we make the body by mechanical processes, each visible particle remains apparently in the same place, however hot it is. The motion which we call heat must therefore be a motion of parts too small to be observed separately; the motions of different parts at the same instant must be in different directions; and the motion of any one part must, at least in solid bodies, be such that, however fast it moves, it never reaches a sensible distance from the point from which it started.

We have now arrived at the conception of a body as consisting of a great many small parts, each of which is in motion. We shall call any one of these parts a molecule of the substance. A molecule may therefore be defined as a small mass of matter the parts of which do not part company during the excursions which the molecule makes when the body to which it belongs is hot.

The doctrine that visible bodies consist of a determinate number of molecules is called the molecular theory of matter. The opposite doctrine is that, however small the parts may be into which we divide a body, each part retains all the properties of the substance. This is the theory of the infinite divisibility of bodies. We do not assert that there is an absolute limit to the divisibility of matter: what we assert is, that after we have divided a body into a certain finite number of constituent parts called molecules, then any further division of these molecules will deprive them of the properties which give rise to the phenomena observed in the substance.

The opinion that the observed properties of visible bodies apparently at rest are due to the action of invisible molecules in rapid motion is to be found in Lucretius.

Daniel Bernoulli was the first to suggest that the pressure of air is due to the impact of its particles on the sides of the vessel containing it; but he made very little progress in the theory which he suggested.

Lesage and Prevost of Geneva, and afterwards Herapath in his 'Mathematical Physics,' made several important applications of the theory.

Dr. Joule in 1848 explained the pressure of gases by the impact of their molecules, and calculated the velocity which they must have to produce the observed pressure.

Krönig also directed attention to this explanation of the phenomena of gases.

It is to Professor Clausius, however, that we owe the recent development of the dynamical theory of gases. Since he took up the subject a great advance has been made by many enquirers. I shall now endeavour to give a sketch of the present state of the theory.

All bodies consist of a finite number of small parts called molecules. Every molecule consists of a definite quantity of matter, which is exactly the same for all the molecules of the same substance. The mode in which the molecule is bound together is also the same for all molecules of the same substance. A molecule may consist of several distinct portions of matter held together by chemical bonds, and may be set in vibration, rotation, or any other kind of relative motion, but so long as the different portions do not part company, but travel together in the excursions made by the molecule, our theory calls the whole connected mass a single molecule.

The molecules of all bodies are in a state of continual agitation. The hotter a body is, the more violently are its molecules agitated. In solid bodies, a molecule, though in continual motion, never gets beyond a certain very small distance from its original position in the body. The path which it describes is confined within a very small region of space.

In fluids, on the other hand, there is no such restriction to the excursions of a molecule. It is true that the molecule generally can travel but a very small distance before its path is disturbed by an encounter with some other molecule; but after this encounter there is nothing which determines the molecule rather to return towards the place from whence it came than to push its way into new regions. Hence in fluids the path of a molecule is not confined within a limited region, as in the case of solids, but may penetrate to any part of the space occupied by the fluid.

The actual phenomena of diffusion both in liquids and in gases furnish the strongest evidence that these bodies consist of molecules in a state of continual agitation.

But when we apply the methods of dynamics to the

investigation of the properties of a system consisting of a great number of small bodies in motion the resemblance of such a system to a gaseous body becomes still more apparent.

I shall endeavour to give some account of what is known of such a system, avoiding all unnecessary mathematical calculations.

ON THE KINETIC THEORY OF GASES.

A gaseous body is supposed to consist of a great number of molecules moving with great velocity. During the greater part of their course these molecules are not acted on by any sensible force, and therefore move in straight lines with uniform velocity. When two molecules come within a certain distance of each other, a mutual action takes place between them, which may be compared to the collision of two billiard balls. Each molecule has its course changed, and starts on a new path. I have concluded from some experiments of my own that the collision between two hard spherical balls is not an accurate representation of what takes place during the encounter of two molecules. A better representation of such an encounter will be obtained by supposing the molecules to act on one another in a more gradual manner, so that the action between them goes on for a finite time, during which the centres of the molecules first approach each other and then separate.

We shall refer to this mutual action as an Encounter between two molecules, and we shall call the course of a molecule between one encounter and another the Free Path of the molecule. In ordinary gases the free motion of a molecule takes up much more time than that occupied by an encounter. As the density of the gas increases, the free path diminishes, and in liquids no part of the course of a molecule can be spoken of as its free path.

In an encounter between two molecules we know that, since the force of the impact acts between the two bodies, the motion of the centre of gravity of the two molecules remains the same after the encounter as it was before. We also know by the principle of the conservation of energy that the velocity of each molecule relatively to the centre of gravity remains the same in magnitude, and is only changed in direction.

Let us next suppose a number of molecules in motion contained in a vessel whose sides are such that if any energy is communicated to the vessel by the encounters of molecules against its sides, the vessel communicates as much energy to other molecules during their encounters with it, so as to preserve the total energy of the enclosed system. The first thing we must notice about this moving system is that even if all the molecules have the same velocity originally, their encounters will produce an inequality of velocity, and that this distribution of velocity will go on continually. Every molecule will then change both its direction and its velocity at every encounter; and, as we are not supposed to keep a record of the exact particulars of every encounter, these changes of motion must appear to us very irregular if we follow the course of a single molecule. If, however, we adopt a statistical view of the system, and distribute the molecules into groups, according to the velocity with which at a given instant they happen to be moving, we shall observe a regularity of a new kind in the proportions of the whole number of molecules which fall into each of these groups.

And here I wish to point out that, in adopting this statistical method of considering the average number of groups of molecules selected according to their velocities, we have abandoned the strict kinetic method of tracing the exact circumstances of each individual molecule in all its encounters. It is therefore possible that we may arrive at results which, though they fairly represent the facts as long as we are supposed to deal with a gas in mass, would cease to be applicable if our faculties and instruments were so sharpened that we could detect and lay hold of each mole cule and trace it through all its course.

For the same reason, a theory of the effects of education deduced from a study of the returns of registrars, in which no names of individuals are given, might be found not to be applicable to the experience of a schoolmaster who is able to trace the progress of each individual pupil.

The distribution of the molecules according to their velocities is found to be of exactly the same mathematical form as the distribution of observations according to the magnitude of their errors, as described in the theory of errors of observation. The distribution of bullet-holes in a target according to their distances from the point aimed at is found to be of the same form, provided a great many shots are fired by persons of the same degree of skill.

We have already met with the same form in the case of heat diffused from a hot stratum by conduction. Whenever in physical phenomena some cause exists over which we have no control, and which produces a scattering of the particles of matter, a deviation of observations from the truth, or a diffusion of velocity or of heat, mathematical expressions of this exponential form are sure to make their appearance.

It appears then that of the molecules composing the system some are moving very slowly, a very few are moving with enormous velocities, and the greater number with intermediate velocities. To compare one such system with another, the best method is to take the mean of the squares of all the velocities. This quantity is called the Mean Square of the velocity. The square root of this quantity is called the Velocity of Mean Square.

DISTRIBUTION OF KINETIC ENERGY BETWEEN TWO DIFFERENT SETS OF MOLECULES.

If two sets of molecules whose mass is different are in motion in the same vessel, they will by their encounters exchange energy with each other till the average kinetic energy of a single molecule of either set is the same. This follows from the same investigation which determines the law of distribution of velocities in a single set of molecules.

Hence if the mass of a molecule of one kind is M_1, and that of a molecule of the other kind is M_2, and if their average velocities of agitation are V_1 and V_2, then

$$M_1 V_1^2 = M_2 V_2^2 \quad . \quad . \quad . \quad . \quad . \quad . \quad . \quad . \quad . \quad (1)$$

The quantity $\frac{1}{2} M V^2$ is called the average kinetic energy of agitation of a single molecule. We shall return to this result when we come to Gay-Lussac's Law of the Volumes of Gases.

INTERNAL KINETIC ENERGY OF A MOLECULE.

If a molecule were a mathematical point endowed with inertia and with attractive and repulsive forces, the only kinetic energy it could possess is that of translation as a whole. But if it be a body having parts and magnitude, these parts may have motions of rotation or of vibration relative to each other, independent of the motion of the centre of gravity of the molecule. We must therefore admit that part of the kinetic energy of a molecule may depend on the relative motions of its parts. We call this the Internal energy, to distinguish it from the energy due to the translation of the molecule as a whole. The ratio of the internal energy to the energy of agitation may be different in different gases.

DEFINITION OF THE VELOCITY OF A GAS.

It is evident that if a gas consists of a great number of molecules moving about in all directions we cannot identify the velocity of any one of these molecules with what we are accustomed to consider as the velocity of the gas itself. Let us consider the case of a gas which has emained in a fixed vessel for a sufficient time to arrive at the normal distribution of velocities. This gas, according to the ordinary notions, is at rest, though the molecules of which it is composed may be flying about in all directions.

Now consider any plane area of an imaginary surface described within the vessel. This surface does not interfere with the motion of the molecules. Some molecules pass through the surface in one direction, and others in the opposite direction; but it is evident, since the gas does not tend to accumulate on one side rather than on the other, that exactly the same number of molecules pass in the one direction as in the other. If, therefore, a gas is at rest, as many molecules pass through a fixed surface in the one direction as in the other in the same time.

It is evident that if the vessel, instead of being at rest, had been in a state of uniform motion, an equal number of molecules would pass in both directions through any surface fixed with respect to the vessel. Hence we find that if a gas is in motion, and if the velocity of a surface coincides in direction and magnitude with that of the gas, the same number of molecules will pass through that surface in the positive direction as in the negative.

This leads to the following definition of the velocity of a gas:

If we determine the motion of the centre of gravity of all the molecules within a very small region surrounding a point in a gas, then the velocity of the gas within that region is defined as the velocity of the centre of gravity of all the molecules within that region.

This is what is meant by the motion of a gas in common language. Besides this motion, there are two other kinds of motion considered in the kinetic theory of gases. The first is the motion of agitation of the molecules. This is the hitherto invisible motion of the molecule considered as a whole. Its course consists of broken portions, called free paths, interrupted by the encounters between different molecules.

The second is the internal motion of each molecule,

consisting partly of rotation and partly of vibrations among the component parts of the molecule.

The velocity of the centre of gravity of a molecule is the resultant of the velocity of the gas and the velocity of agitation of the individual molecule at the given instant. The velocity of a constituent part of a molecule is the resultant of the velocity of its centre of gravity and the velocity of the constituent part relatively to the centre of gravity of the molecule.

THEORY OF THE PRESSURE OF A GAS.

Let us consider two portions of a gas separated by a plane surface which moves with the same velocity as the gas. We have seen that in this case the number of molecules which pass through the plane in opposite directions is the same.

FIG. 41.

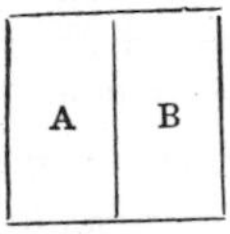

Each molecule in crossing the plane from the region A to the region B enters the second region in precisely the same state as it leaves the first. It therefore carries over into the region B, not only its mass, but its momentum and its kinetic energy. Hence, if we consider the quantity of momentum in a given direction existing at any instant in the particles in the region B, this quantity will be altered whenever a molecule crosses the boundary, carrying its momentum along with it.

Now let us consider all the molecules whose velocity differs by less than a certain quantity, c, from a given velocity the components of which are u in the direction perpendicular to the plane from A towards B, and v and w in two other directions parallel to the plane. Let there be N molecules whose velocity is within these limits in every unit of volume, and let the mass of each of these be M.

Then the number of these molecules which will cross unit of area of the plane from A to B in unit of time is

$$\text{N}\, u.$$

The momentum of each of these molecules resolved in the direction A B is M u.

Hence the momentum in this direction communicated to the region B in unit of time is

$$\text{M N } u^2.$$

Since this bombardment of the region B does not produce motion of the gas, a pressure must be exerted on the gas by the sides of the vessel, and the amount of this pressure for every unit of area must be M N u^2.

The region A loses positive momentum at the same rate, and in order to preserve equilibrium there must be a pressure equal to M N u^2 on every unit of area of the surface of the region A.

Hitherto we have considered only one group of molecules, whose velocities lie between given limits. In every such group that which determines the pressure in the direction A B on the surface separating A from B is a quantity of the form M N u^2, where N is the number of molecules in the group, and u is the velocity of each molecule resolved in the direction A B. The other components of the velocity do not influence the pressure in this direction.

To find the whole pressure, we must find the sum of all such expressions as M N u^2 for all the groups of molecules in the system. We may write this result p = M N $\overline{u^2}$, where N now signifies the total number of molecules in unit of volume, and $\overline{u^2}$ denotes the mean value of u^2 for all these molecules. Now if V^2 is the square of the velocity without regard to direction,

$$\text{V}^2 = u^2 + v^2 + w^2,$$

where u v w are the components in three directions at right angles. Hence if $\overline{u^2}$, $\overline{v^2}$, and $\overline{w^2}$ denote the mean square of these components, and $\overline{\text{V}^2}$ the mean square of the resultant,

$$\overline{\text{V}^2} = \overline{u^2} + \overline{v^2} + \overline{w^2}.$$

When, as in every gas at rest, the pressure is equal in all directions, $\overline{u^2} = \overline{v^2} = \overline{w^2}$, and therefore $\overline{V^2} = 3\,\overline{u^2}$.

Hence the pressure of a gas is

$$p = \tfrac{1}{3} M N \overline{V^2} \quad . \quad . \quad . \quad . \quad . \quad . \quad . \quad . \quad . \quad . \quad (2)$$

where M is the mass of each molecule, N is the number of molecules in unit of volume, and $\overline{V^2}$ is the mean square of the velocity.

In this expression there are two quantities which have never been directly measured—the mass of a single molecule, and the number of molecules in unit of volume. But we have here to do with the product of these quantities, which is evidently the mass of the substance in unit of volume, or, in other words, its density. Hence we may write the expression

$$p = \tfrac{1}{3} \rho \overline{V^2} \quad . \quad . \quad . \quad . \quad . \quad . \quad . \quad . \quad . \quad . \quad (3)$$

where ρ is the density of the gas.

It is easy from this expression to determine, as was first done by Joule, the mean square of the velocity of the molecules of a gas, for

$$\overline{V^2} = 3 \frac{p}{\rho} \quad . \quad . \quad . \quad . \quad . \quad . \quad . \quad . \quad . \quad . \quad . \quad (4)$$

where p is the pressure, and ρ the density, which must of course be expressed in terms of the same fundamental units.

For instance, under the atmospheric pressure of 2116·4 pounds weight on the square foot, and at the temperature of melting ice, the density of hydrogen is 0·005592 pounds in a cubic foot. Hence $\frac{p}{\rho} = 378816$ in gravitation units, and if the intensity of gravity where this relation was observed was 32·2, we have

$$\overline{V^2} = 36593916,$$

or, taking the square root of this quantity,

$$\overline{V} = 6097 \text{ feet per second.}$$

This is the velocity of mean square for the molecules of hydrogen at 32° F. and at the atmospheric pressure.

LAW OF BOYLE.

Two bodies are said to be of the same temperature when there is no more tendency for heat to pass from the first to the second than in the reverse direction. In the kinetic theory of heat, as we have seen, this thermal equilibrium is established when there is a certain relation between the velocities of agitation of the molecules of the two bodies. Hence the temperature of a gas must depend on the velocity of agitation of its molecules, and this velocity must be the same at the same temperature, whatever be the density.

In the expression $p = \frac{1}{3} \rho \bar{V}^2$, the quantity $\bar{V}^2$ depends only on the temperature as long as the gas remains the same. Hence when the density ρ varies, the pressure p must vary in the same proportion. This is Boyle's law, which is now raised from the rank of an experimental fact to that of a deduction from the kinetic theory of gases.

If v denotes the volume of unit of mass, we may write this expression

$$p v = \frac{1}{3} \bar{V}^2 \quad . \quad . \quad . \quad . \quad . \quad . \quad . \quad . \quad . \quad (5)$$

Now $p v$ is proportional to the absolute temperature, as measured by a thermometer, of the particular gas under consideration. Hence $\bar{V}^2$, the mean square of the velocity of agitation, is proportional to the absolute temperature measured in this way.

LAW OF GAY-LUSSAC.

Let us next consider two different gases in thermal equilibrium. We have already stated that if M_1 M_2 are the masses of individual molecules of these gases, and V_1 V_2 their respective velocities of agitation, it is necessary for thermal equilibrium that $M_1 \bar{V}_1^2 = M_2 \bar{V}_2^2$ by equation (1).

If the pressures of these gases are p_1 and p_2, and the number of molecules in unit of volume N_1 and N_2, then, by equation (2),

$$p_1 = \frac{1}{3} M_1 N_1 \bar{V}_1^2 \text{ and } p_2 = \frac{1}{3} M_2 N_2 \bar{V}_2^2.$$

If the pressures of the two gases are equal,

$$M_1 N_1 \bar{V}_1^2 = M_2 N_2 \bar{V}_2^2.$$

If their temperatures are equal,

$$M_1 \bar{V}_1^2 = M_2 \bar{V}_2^2.$$

Dividing the terms of the first of these equations by those of the second, we find

$$N_1 = N_2 \text{ . . . . . } \quad (6)$$

or *when two gases are at the same pressure and temperature, the number of molecules in unit of volume is the same in both gases.*

If we put $\rho_1 = M_1 N_1$ and $\rho_2 = M_2 N_2$ for the densities of the two gases, then, since $N_1 = N_2$, we get

$$\rho_1 : \rho_2 :: M_1 : M_2 \text{ . . . . . . } \quad (7)$$

or *the densities of two gases at the same temperature and pressure are proportional to the masses of their individual molecules.*

These two equivalent propositions are the expression of a very important law established by Gay-Lussac, that the densities of gases are proportional to their molecular weights.

The proportion by weight in which different substances combine to form chemical compounds depends, according to Dalton's atomic theory, on the weights of their molecules, and it is one of the most important researches in chemistry to determine the proportions of the weights of the molecules from the proportions in which they enter into combination. Gay-Lussac discovered that in the case of gases the volumes of the combining quantities of different gases always stand in a simple ratio to each other. This law of volumes has now been raised from the rank of an empirical fact to that of a deduction from our theory, and we may now assert, as a dynamical proposition, that the weights of the molecules of gases (that is, those small portions which do not part company during their motion) are proportional to the densities of these gases at standard temperature and pressure.

LAW OF CHARLES.

We must next consider the effect of changes of temperature on different gases. Since at all temperatures, when there is thermal equilibrium,

$$M_1 \overline{V_1}^2 = M_2 \overline{V_2}^2;$$

and since the absolute temperature, as measured by a gas thermometer, is proportional to $\overline{V_1}^2$ when the gas is of the first kind, and to $\overline{V_2}^2$ when the gas is of the second kind; it follows, since $\overline{V_1}^2$ is itself proportional to $\overline{V_2}^2$, that the absolute temperatures, as measured by the two thermometers, are proportional, and if they agree at any one temperature (as the freezing point), they agree throughout. This is the law of the equal dilatation of gases discovered by Charles.

KINETIC ENERGY OF A MOLECULE.

The mean kinetic energy of agitation of a molecule is the product of its mass by half the mean square of its velocity, or

$$\tfrac{1}{2} M \overline{V^2}.$$

This is the energy due to the motion of the molecule as a whole, but its parts may be in a state of relative motion. If we assume, with Clausius, that the energy due to this internal motion of the parts of the molecule tends towards a value having a constant ratio to the energy of agitation, the whole energy will be proportional to the energy of agitation, and may be written

$$\tfrac{1}{2} \beta M \overline{V^2},$$

where β is a factor, always greater than unity, and probably equal to 1·634 for air and several of the more perfect gases. For steam it may be as much as 2·19, but this is very uncertain.

To find the kinetic energy of the substance contained in unit of volume, we have only to multiply by the number of molecules, and we obtain

$$T = \tfrac{1}{2} \beta M N \bar{V}^2 \quad . \quad . \quad . \quad . \quad . \quad . \quad . \quad . \quad . \quad . \quad (8)$$

Comparing this with the equation (2) which determines the pressure, we get

$$T_v = \tfrac{3}{2} \beta p \quad . \quad . \quad . \quad . \quad . \quad . \quad . \quad . \quad . \quad . \quad (9)$$

or the energy in unit of volume is numerically equal to the pressure on unit of area multiplied by $\frac{3}{2}\beta$.

The energy in unit of mass is found by multiplying this by v, the volume of unit of mass :

$$T_m = \tfrac{3}{2} \beta p v \quad . \quad . \quad . \quad . \quad . \quad . \quad . \quad . \quad . \quad . \quad (10)$$

SPECIFIC HEAT AT CONSTANT VOLUME.

Since the product $p v$ is proportional to the absolute temperature, the energy is proportional to the temperature.

The specific heat is measured dynamically by the increase of energy corresponding to a rise of one degree of temperature. Hence

$$K_v = \tfrac{3}{2} \beta \frac{p v}{\theta} \quad . \quad . \quad . \quad . \quad . \quad . \quad . \quad . \quad . \quad (11)$$

To express the specific heat in ordinary thermal units, we must divide this by J, the specific heat of water (Joule's equivalent). It follows from this expression that for any one gas the specific heat of unit of mass at constant volume is the same for all pressures and temperatures, because $\frac{p v}{\theta}$ remains constant. For different gases the specific heat at constant volume is inversely proportional to the specific gravity, and directly proportional to β.

Since β is nearly the same for several gases, the specific heat of these gases is inversely proportional to their specific gravity referred to air, or, since the specific gravity is proportional to their molecular weight, the specific heat multiplied by the molecular weight is the same for all these gases.

This is the law of Dulong and Petit. It would be accurate for all gases if the value of β were the same in every case.

It has been shown at p. 181 that the difference of the two specific heats is $\frac{p\,v}{\theta}$. Hence their ratio, γ, is

$$\gamma = \frac{2}{3\,\beta} + 1 \quad \text{and} \quad \beta = \tfrac{2}{3}\,\frac{1}{\gamma - 1}.$$

If U is the velocity of sound in a gas, we have, as at p. 208,

$$U^2 = \gamma\, p\, v \quad . \quad . \quad . \quad . \quad . \quad . \quad . \quad . \quad . \quad . \quad (12)$$

The mean square of the velocity of agitation is

$$\overline{V}^2 = 3\,p\,v \quad . \quad . \quad . \quad . \quad . \quad . \quad . \quad . \quad . \quad . \quad (13)$$

Hence $U = \sqrt{\frac{\gamma}{3}}\,V$, or, if $\gamma = 1{\cdot}408$, as in air and several other gases,

$$U = {\cdot}6858\ V \quad \text{or} \quad V = 1{\cdot}458\ U \quad . \quad . \quad (14)$$

These are the relations between the velocity of sound and the velocity of mean square of agitation in any gas for which $\gamma = 1{\cdot}408$.

The nature of this book admits only of a brief account of some other results of the kinetic theory of gases. Two of these are independent of the nature of the action between the molecules during their encounters.

The first of these relates to the equilibrium of a mixture of gases acted on by gravity. The result of our theory is that the final distribution of any number of kinds of gas in a vertical vessel is such that the density of each gas at a given height is the same as if all the other gases had been removed, leaving it alone in the vessel.

This is exactly the mode of distribution which Dalton supposed to exist in a mixed atmosphere in equilibrium, the law of diminution of density of each constituent gas being the same as if no other gases were present.

In our atmosphere the continual disturbances caused by winds carry portions of the mixed gases from one stratum to another, so that the proportion of oxygen and nitrogen at different heights is much more uniform than if these gases had been allowed to take their places by diffusion during a dead calm.

The second result of our theory relates to the thermal equilibrium of a vertical column. We find that if a vertical column of a gas were left to itself, till by the conduction of heat it had attained a condition of thermal equilibrium, the temperature would be the same throughout, or, in other words, gravity produces no effect in making the bottom of the column hotter or colder than the top.

This result is important in the theory of thermodynamics, for it proves that gravity has no influence in altering the conditions of thermal equilibrium in any substance, whether gaseous or not. For if two vertical columns of different substances stand on the same perfectly conducting horizontal plate, the temperature of the bottom of each column will be the same ; and if each column is in thermal equilibrium of itself, the temperatures at all equal heights must be the same. In fact, if the temperatures of the tops of the two columns were different, we might drive an engine with this difference of temperature, and the refuse heat would pass down the colder column, through the conducting plate, and up the warmer column; and this would go on till all the heat was converted into work, contrary to the second law of thermodynamics.

But we know that if one of the columns is gaseous, its temperature is uniform. Hence that of the other must be uniform, whatever its material.

This result is by no means applicable to the case of our atmosphere. Setting aside the enormous direct effect of the sun's radiation in disturbing thermal equilibrium, the effect of winds in carrying large masses of air from one height to another tends to produce a distribution of temperature of a quite different kind, the temperature at any height being such that a mass of air, brought from one height to another without gaining or losing heat, would always find itself at the temperature of the surrounding air. In this condition of what Sir William Thomson has called the Convective equilibrium of heat, it is not the temperature which is constant, but the quantity ϕ, which determines the adiabatic curves.

In the convective equilibrium of temperature, the absolute temperature is proportional to the pressure raised to the power $\frac{\gamma - 1}{\gamma}$, or 0·233.

The extreme slowness of the conduction of heat in air, compared with the rapidity with which large masses of air are carried from one height to another by the winds, causes the temperature of the different strata of the atmosphere to depend far more on this condition of convective equilibrium than on true thermal equilibrium.

We now proceed to those phenomena of gases which, according to the kinetic theory, depend upon the particular nature of the action which takes place when the molecules encounter each other, and on the frequency of these encounters.

There are three phenomena of this kind of which the kinetic theory takes account—the diffusion of gases, the viscosity of gases, and the conduction of heat through a gas.

We have already described the known facts about the interdiffusion of two different gases. It is only when the gases are chemically different that we can trace the process of diffusion, but on the molecular theory diffusion is always going on, even in a single gas; only it is impossible to trace the progress of the molecules, because we cannot tell one from another.

The relation between diffusion and viscosity may be explained as follows: Consider the case of motion of a mass of gas, which has already been described in Chapter XXI., in which the different horizontal layers of the gas slide over each other. In diffusion the molecules pass, some of them upwards and some of them downwards, through any horizontal plane. If the medium has different properties of any kind above and below this plane, then this interchange of molecules will tend to assimilate the properties of the two portions of the medium.

In the case of ordinary diffusion, the proportions of the two diffusing substances are different above and below, and vary in the different horizontal layers according to their height. In the case of internal friction, the mean horizontal momentum is different in the different layers, and when the molecules pass through the plane, carrying their momentum with them, this exchange of momentum between the upper and lower parts of the medium constitutes a force tending to equalize their velocity, and this is the phenomenon actually observed in the motion of viscous fluids.

The coefficient of viscosity, when measured in the kinematic way, represents the rate at which the equalization of velocity goes on by the exchange of the momentum of the molecules, just as the coefficient of diffusion represents the rate at which the equalization of chemical composition goes on by the exchange of the molecules themselves.

It appears from the kinetic theory of gases that if D is the coefficient of diffusion of the gas *into itself*, and ν the viscosity measured kinematically,

$$\nu = 0\cdot6479\, D \quad \text{(15)}$$

$$D = 1\cdot5435\, \nu \quad \text{(16)}$$

The conduction of heat in a gas, according to the kinetic theory, is simply the diffusion of the energy of the molecules by their moving about in the medium and carrying their energy with them till they encounter other molecules, when the energy is redistributed. The relation of the conductivity κ, measured thermometrically, to the viscosity ν, measured kinematically, is

$$\kappa = \frac{5}{3\gamma}\nu \quad \text{(17)}$$

It appears, therefore, that diffusion, viscosity, and conductivity in gases are related to each other in a very simple way, being the rate of equalization of three properties of the medium—the proportion of its ingredients, its velocity, and its temperature. The equalization is effected by the same agency in each case—namely, the agitation of the molecules. In each case, if the density remains the same, the rate of equalization is proportional to the absolute temperature; and if the temperature remains the same, the rate of equalization is inversely proportional to the density. Hence, if we consider the temperature and the pressure as defining the state of the gas, the quantities D, ν, and κ vary directly as the square of the absolute temperature and inversely as the pressure.

MOLECULAR THEORY OF EVAPORATION AND CONDENSATION.

The mathematical difficulties arising in the investigation of the motions of molecules are so great that it is not to be wondered at that most of the numerical results are confined to the phenomena of gases. The general character, however, of the explanation of many other phenomena by the molecular theory has been pointed out by Clausius and others.

We have seen that in the case of a gas some of the molecules have a much greater velocity than others, so that it is only to the average velocity of all the molecules that we can ascribe a definite value. It is probable that this is also true of the motions of the molecules of a liquid, so that, though the average velocity may be much smaller than in the vapour of that liquid, some of the molecules in the liquid may have velocities equal to or greater than the average velocity in the vapour. If any of the molecules at the surface of the liquid have such velocities, and if they are moving *from* the liquid, they will escape from those forces which retain the other molecules as constituents of the liquid, and will fly about as vapour in the space outside the liquid. This is the molecular theory of evaporation. At the same time, a molecule of the vapour striking the liquid may become entangled among the molecules of the liquid, and may thus become part of the liquid. This is the molecular explanation of condensation. The number of molecules which pass from the liquid to the vapour depends on the temperature of the liquid. The number of molecules which pass from the vapour to the liquid depends upon the density of the vapour as well as its temperature. If the temperature of the vapour is the same as that of the liquid, evaporation will take place as long as more molecules are evaporated than condensed; but when the density of the vapour has increased to such a value that as many molecules are condensed as evaporated, then the vapour has attained its maximum density. It is then said to be saturated, and it is commonly supposed that evaporation ceases. According to the molecular theory, however, evaporation is still going on as fast as ever; only, condensation is also going on at an equal rate, since the proportions of liquid and of gas remain unchanged.

We have hitherto spoken of the case in which the liquid and the vapour consist of the same substance in different states.

A similar explanation, however, applies to cases in which the vapour or gas is absorbed by a liquid of a different kind, as when oxygen or carbonic acid is absorbed by water or alcohol. In such cases a 'movable equilibrium' is attained when the liquid has absorbed a quantity of the gas whose volume at the density of the unabsorbed gas is a certain multiple or fraction of the volume of the liquid; or, in other words, the density of the gas in the liquid and outside the liquid stand in a certain numerical ratio to each other. This subject is treated very fully in Bunsen's 'Gasometry.'

The amount of vapour of a liquid diffused into a gas of a different kind is generally independent of the nature of the gas, except when the gas acts chemically on the vapour.

MOLECULAR THEORY OF ELECTROLYSIS.

A very interesting part of molecular science which has not been thoroughly worked out, but which hardly belongs to a treatise on Heat, is the theory of electrolysis. Here an electromotive force acting on a liquid electrolyte causes the molecules of one of its components to be urged in one direction, while those of the other component are urged in the opposite direction. Now these components are joined together in pairs by chemical forces of great power, so that we might expect that no electrolytic effect could take place unless the electromotive force were so strong as to be able to tear these couples asunder. But, according to Clausius, in the dance of molecules which is always going on, some of the linked pairs of molecules acquire such velocities that when they have an encounter with a pair also in violent motion the molecules composing one or both of the pairs are torn asunder, and wander about seeking new partners. If the temperature is so high that the general agitation is so violent that more pairs of molecules are torn asunder than can pair again in an equal time, we have the phenomenon of Dissociation, studied by M. Ste.-Claire Deville. If, on the other hand, the separated molecules can always find partners before they are ejected from the system, the composition of the system remains apparently the same.

Now Professor Clausius considers that it is during these temporary separations that the electromotive force comes into play as a directing power, causing the molecules of one component to move on the whole one way, and those of the other the opposite way. Thus the component molecules are always changing partners, even when no electromotive force is in action, and the only effect of this force is to give direction to those movements which are already going on.

Professor Wiedemann, who has also taken this view of electrolysis, compares the phenomenon with that of diffusion, and shows that the electric conductivity of an electrolyte may be considered as depending on the coefficient of diffusion of the components through each other.

MOLECULAR THEORY OF RADIATION.

The phenomena already described are explained on the molecular theory by the motion of agitation of the molecules, a motion which is exceedingly irregular, the intervals between successive encounters and the velocities of a molecule during successive free paths not being subject to any law which we can express. The internal motion of a single molecule is of a very different kind. If the parts of the molecule are capable of relative motion without being altogether torn asunder, this relative motion will be some kind of vibration. The small vibrations of a connected system may be resolved into a number of simple vibrations, the law of each of which is similar to that of a pendulum. It is probable that in gases the molecules may execute many of such vibrations in the interval between successive encounters. At each encounter the whole molecule is roughly shaken. During its free path it vibrates according to its own laws, the amplitudes of the different simple vibrations being determined by the nature of the collision, but their periods depending only on the constitution of the molecule itself. If the molecule is capable of communicating these vibrations to the medium in which radiations are propagated, it will send forth radiations of certain definite kinds, and if these belong to the luminous part of the spectrum, they will be visible as light of definite refrangibility. This, then, is the explanation, on the molecular theory, of the bright lines observed in the spectra of incandescent gases. They represent the disturbance communicated to the luminiferous medium by molecules vibrating in a regular and periodic manner during their free paths. If the free path is long, the molecule, by communicating its vibrations to the ether, will cease to vibrate till it encounters some other molecule.

By raising the temperature we increase the velocity of

the motion of agitation and the force of each encounter. The higher the temperature the greater will be the amplitude of the internal vibrations of all kinds, and the more likelihood will there be that vibrations of short period will be excited, as well as those fundamental vibrations which are most easily produced. By increasing the density we diminish the length of the free path of each molecule, and thus allow less time for the vibrations excited at each encounter to subside, and, since each fresh encounter disturbs the regularity of the series of vibrations, the radiation will no longer be capable of complete resolution into a series of vibrations of regular periods, but will be analysed into a spectrum showing the bright bands due to the regular vibrations, along with a ground of diffused light, forming a continuous spectrum due to the irregular motion introduced at each encounter.

Hence when a gas is rare the bright lines of its spectrum are narrow and distinct, and the spaces between them are dark. As the density of the gas increases, the bright lines become broader and the spaces between them more luminous.

There is another reason for the broadening of the bright lines and the luminosity of the whole spectrum in dense gases, which we have already stated at p. 225. There is this difference, however, between the effect there mentioned and that described here. At p. 225 the light from a certain stratum of incandescent gas was supposed to penetrate through other strata, which absorbed the brighter rays faster than the less luminous ones. This effect depends only on the total quantity of gas through which the rays pass, and will be the same whether it is a mile of gas at thirty inches pressure, or thirty miles at one inch pressure. The effect which we are now considering depends on the absolute density, so that it is by no means the same whether a stratum containing a given quantity of gas is one mile or thirty miles thick.

When the gas is so far condensed that it assumes the liquid or solid form, then, as the molecules have no free path, they have no regular vibrations, and no bright lines are commonly observed in incandescent liquids or solids. Mr. Huggins, however, has observed bright lines in the spectrum of incandescent erbia and lime, which appear to be due to the solid matter, and not to its vapour.

LIMITATION OF THE SECOND LAW OF THERMODYNAMICS.

Before I conclude, I wish to direct attention to an aspect of the molecular theory which deserves consideration.

One of the best established facts in thermodynamics is that it is impossible in a system enclosed in an envelope which permits neither change of volume nor passage of heat, and in which both the temperature and the pressure are everywhere the same, to produce any inequality of temperature or of pressure without the expenditure of work. This is the second law of thermodynamics, and it is undoubtedly true as long as we can deal with bodies only in mass, and have no power of perceiving or handling the separate molecules of which they are made up. But if we conceive a being whose faculties are so sharpened that he can follow every molecule in its course, such a being, whose attributes are still as essentially finite as our own, would be able to do what is at present impossible to us. For we have seen that the molecules in a vessel full of air at uniform temperature are moving with velocities by no means uniform, though the mean velocity of any great number of them, arbitrarily selected, is almost exactly uniform. Now let us suppose that such a vessel is divided into two portions, A and B, by a division in which there is a small hole, and that a being, who can see the individual molecules, opens and closes this hole, so as to allow only the swifter molecules to pass from A to B, and only the slower ones to pass from B to A. He will thus, without expenditure of work, raise the temperature of B and lower that of A, in contradiction to the second law of thermodynamics.

This is only one of the instances in which conclusions which we have drawn from our experience of bodies consisting of an immense number of molecules may be found not to be applicable to the more delicate observations and experiments which we may suppose made by one who can perceive and handle the individual molecules which we deal with only in large masses.

In dealing with masses of matter, while we do not perceive the individual molecules, we are compelled to adopt what I have described as the statistical method of calculation, and to abandon the strict dynamical method, in which we follow every motion by the calculus.

It would be interesting to enquire how far those ideas about the nature and methods of science which have been derived from examples of scientific investigation in which the dynamical method is followed are applicable to our actual knowledge of concrete things, which, as we have seen, is of an essentially statistical nature, because no one has yet discovered any practical method of tracing the path of a molecule, or of identifying it at different times.

I do not think, however, that the perfect identity which we observe between different portions of the same kind of matter can be explained on the statistical principle of the stability of the averages of large numbers of quantities each of which may differ from the mean. For if of the molecules of some substance such as hydrogen, some were of slightly greater mass than others, we have the means of producing a separation between molecules of different masses, and in this way we should be able to produce two kinds of hydrogen, one of which would be somewhat denser than the other. As this cannot be done, we must admit that the equality which we assert to exist between the molecules of hydrogen applies to each individual molecule, and not merely to the average of groups of millions of molecules.

NATURE AND ORIGIN OF MOLECULES.

We have thus been led by our study of visible things to a theory that they are made up of a finite number of parts or molecules, each of which has a definite mass, and possesses other properties. The molecules of the same substance are all exactly alike, but different from those of other substances. There is not a regular gradation in the mass of molecules from that of hydrogen, which is the least of those known to us, to that of bismuth ; but they all fall into a limited number of classes or species, the individuals of each species being exactly similar to each other, and no intermediate links are found to connect one species with another by a uniform gradation.

We are here reminded of certain speculations concerning the relations between the species of living things. We find that in these also the individuals are naturally grouped into species, and that intermediate links between the species are wanting. But in each species variations occur, and there is a perpetual generation and destruction of the individuals of which the species consist.

Hence it is possible to frame a theory to account for the present state of things by means of generation, variation, and discriminative destruction.

In the case of the molecules, however, each individual is permanent ; there is no generation or destruction, and no variation, or rather no difference, between the individuals of each species.

Hence the kind of speculation with which we have become so familiar under the name of theories of evolution is quite inapplicable to the case of molecules.

It is true that Descartes, whose inventiveness knew no bounds, has given a theory of the evolution of molecules. He supposes that the molecules with which the heavens are nearly filled have received a spherical form from the long-continued grinding of their projecting parts, so that, like marbles in a mill, they have 'rubbed each other's angles down.' The result of this attrition forms the finest kind of molecules, with which the interstices between the globular molecules are filled. But, besides these, he describes another elongated kind of molecules, the *particula striata*, which have received their form from their often threading the interstices between three spheres in contact. They have thus acquired three longitudinal ridges, and, since some of them during their passage are rotating on their axes, these ridges are not in general parallel to the axis, but are twisted like the threads of a screw. By means of these little screws he most ingeniously attempts to explain the phenomena of magnetism.

But it is evident that his molecules are very different from ours. His seem to be produced by some general break-up of his solid space, and to be ground down in the course of ages, and, though their relative magnitude is in some degree determinate, there is nothing to determine the absolute magnitude of any of them.

Our molecules, on the other hand, are unalterable by any of the processes which go on in the present state of things, and every individual of each species is of exactly the same magnitude, as though they had all been cast in the same mould, like bullets, and not merely selected and grouped according to their size, like small shot.

The individuals of each species also agree in the nature of the light which they emit—that is, in their natural periods of vibration. They are therefore like tuning-forks all tuned to concert pitch, or like watches regulated to solar time.

In speculating on the cause of this equality we are debarred from imagining any cause of equalization, on account of the immutability of each individual molecule. It is difficult, on the other hand, to conceive of selection and elimination of intermediate varieties, for where can these eliminated molecules have gone to if, as we have reason to believe, the hydrogen, &c., of the fixed stars is composed of molecules identical in all respects with our own? The time required to eliminate from the whole of the visible universe every molecule whose mass differs from that of some one of our so-called elements, by processes similar to Graham's method of dialysis, which is the only method we can conceive of at present, would exceed the utmost limits ever demanded by evolutionists as many times as these exceed the period of vibration of a molecule.

But if we suppose the molecules to be made at all, or if we suppose them to consist of something previously made, why should we expect any irregularity to exist among them? If they are, as we believe, the only material things which still remain in the precise condition in which they first began to exist, why should we not rather look for some indication of that spirit of order, our scientific confidence in which is never shaken by the difficulty which we experience in tracing it in the complex arrangements of visible things, and of which our moral estimation is shown in all our attempts to think and speak the truth, and to ascertain the exact principles of distributive justice?

APPENDIX.

Table of the Coefficients of Interdiffusion of Gases, from the Memoir of Professor Loschmidt (see p. 259), in square centimètres per second.

		D
Carbonic acid	Air	·1423
—	Hydrogen	·5614
—	Oxygen	·1409
—	Marsh gas	·1586
—	Carbonic oxide	·1406
—	Nitrous oxide	·0982
Oxygen	Hydrogen	·7214
—	Carbonic oxide	·1802
Carbonic oxide	Hydrogen	·6422
Sulphurous acid	Hydrogen	·4800

Professor J. Stefan, also of Vienna, has undertaken a series of very delicate experiments to determine the thermal conductivity of air and other gases. He finds the thermometric conductivity, κ, of air 0·256 square centimètres per second. The rate of propagation of thermal effects in still air is therefore intermediate between the rate in iron, for which $\kappa = 0{\cdot}183$, and in copper, for which $\kappa = 1{\cdot}077$. Stefan finds it intermediate between iron and zinc.

The calorimetric conductivity, k, is 0·0000558 for air, or about 20,000 times less than that of copper, and 3,360 times less than that of iron. As calculated from the coefficient of viscosity by the writer $k = 0{\cdot}000054$.

Stefan has also found that the calorimetric conductivity is independent of the pressure, and that it is seven times greater for hydrogen than for air. Both these results had been predicted by the molecular theory. See Maxwell 'On the Dynamical Theory of Gases,' *Phil. Trans.* 1867, p. 88.